T0338247

The Environment and Landscape in Motorway Design

The Environment and Landscape in Motorway Design

Qian Guochao
Deputy Director
Jiangsu Provincial Department of Transportation
People's Republic of China

Tang Shuyu
Researcher
Institute of Botany,
Jiangsu Province and Chinese Academy of Sciences
People's Republic of China

Zhao Min
Senior Engineer and Project Department Director
Jiangsu Provincial Expressway Construction Headquarters
People's Republic of China

Jing Chun
Senior Engineer
Jiangsu Provincial Expressway Construction Headquarters
People's Republic of China

WILEY Blackwell

China Communications Press

This edition first published 2014
Registered office
John Wiley & Sons, Ltd, The Atrium, Southern Gate, Chichester, West Sussex, PO19 8SQ,
United Kingdom.

Editorial offices:
9600 Garsington Road, Oxford, OX4 2DQ, United Kingdom.
The Atrium, Southern Gate, Chichester, West Sussex, PO19 8SQ, United Kingdom.

For details of our global editorial offices, for customer services and for information about how to
apply for permission to reuse the copyright material in this book please see our website at
www.wiley.com/wiley-blackwell.

Originally published in the Chinese language by China Communications Press, Beijing,
China 100011, as 高速公路环境景观设计.
© 2012 China Communications Press

Library of Congress Cataloging-in-Publication Data

Guochao, Qian.
  The environment and landscape in highway design / Qian Guochao, Tang Shuyu, Zhao Min,
Jing Chun.
       pages cm
  Includes bibliographical references and index.
  ISBN 978-1-118-33297-9 (cloth)
  1. Roadside improvement. 2. Roads–Environmental aspects. I. Shuyu, Tang. II. Min, Zhao.
III. Chun, Jing. IV. Title.
  TE177.G87 2014
  713–dc23
                                                                              2014005801

A catalogue record for this book is available from the British Library.

Wiley also publishes its books in a variety of electronic formats. Some content that appears in
print may not be available in electronic books.

Cover images courtesy of the authors and iStockphoto
Cover design by Steve Thompson

Set in 9.5/13 in MinionPro by Laserwords Private Limited, Chennai, India
Printed and bound in Malaysia by Vivar Printing Sdn Bhd

1   2014

Contents

About the Authors

Qian Guochao is a professor-level Senior Engineer, formerly Deputy Commander of Jiangsu Provincial Expressway Construction Headquarters, and is now Deputy Director of Jiangsu Provincial Department of Transportation, whose major research fields cover expressways and bridges.

Tang Shuyu is a Researcher at the Institute of Botany, Jiangsu Province and Chinese Academy of Sciences, and Chief Engineer of Nanjing Botanical Garden Mem. Sun Yat-Sen, China Academy of Sciences, whose major research fields cover ecological landscapes and gardens.

Zhao Min is a professor-level Senior Engineer and Project Department Director of Jiangsu Provincial Expressway Construction Headquarters, whose major research fields cover expressways and bridges.

Jing Chun is a Senior Engineer at the Jiangsu Provincial Expressway Construction Headquarters, whose major research fields cover expressways and bridges.

With Contributions from

Zheng Chenhui
Zhu Juhui
Lu Jianguo
Jiang Peng
Tang Ren
Xu Chi
Zhang Zhaowu
Li Yunlong
Hao Riming
Lü Weiguo
Zhang Song
Wang Zheng
Sun Haijun

Introduction

Since the world's first motorway was completed in Germany in 1932, more than 80 other countries have built motorways, with a total length of more than 220 000 km.

European countries started quite early in the construction of motorways and made rapid progress in connecting motorways between cities and countries into a complete network. A well-developed motorway network and a smooth and efficient driving environment facilitate convenient access to these countries.

For motorway construction, European countries have paid special attention to environmental design. Full consideration is given not only to the layout, but also to whether the motorways can be integrated into the natural environment and landscape. In addition, road alignment and the visual and psychological impact on drivers and passengers are also considered in design. At present, these countries have made great achievements in terms of ecological protection and the use of information resources when constructing motorways. The environmental landscape is also very picturesque.

Motorway landscape design began in the early 1920s. Specialized landscape design was emphasized in the construction of Parkway in the USA, mainly focusing on how the motorway alignment would fit into the natural landscape as well as the protection and utilization of scenery along the route. In the 1930s, Germany first adopted an alignment model to check and correct space alignment and came up with an integrated design taking into consideration horizontal and vertical factors, thereby achieving optimum motorway design. Further progress was later made in motorway landscape design, such as using a customized model and plastic foamboard to design a route model and making perspective drawings using the optical projection principle.

Since the 1960s, many developed countries have begun to consider landscape design in motorway construction while attaching importance to the improvement of existing motorway landscapes, and these countries have also developed appropriate specifications and regulations. For example, in 1965 the United States issued the *Highway Beautification Act*, and *Development Guidelines of U.S. Interstate and Defense Highway Landscape* and *Guidelines for*

The Environment and Landscape in Motorway Design, First Edition.
Qian Guochao, Tang Shuyu, Zhao Min and Jing Chun.
© 2014 China Communications Press. Published 2014 by John Wiley & Sons, Ltd.

Highway Landscape; later, in 1970 they formulated *Guidelines for Highway Landscape and Environmental Design* and *Practical Highway Aesthetics*. All these specifications and regulations focus on visually attractive highways with basic functions which aim to be harmonious with the surroundings. They also published some basic principles, for example the diversity of the landscape along the route must be ensured in motorway design and the whole route should be scenic; the motorway must 'adapt to the terrain' without large-scale cutting and filling; harmony between the motorway and the surrounding landscape must be realized as far as possible; and the natural landscape must revert to its original state rapidly, or the natural appearance must be restored by appropriate planting and greening if damage to the natural landscape is inevitable in construction. Meanwhile, basic principles of landscape design have also been broadly adopted in the construction of motorways, trunk highways as well as scenic highways in other developed countries such as Germany, France, the UK and Japan. Regulations governing landscape design have been developed in relevant design specifications. In 1974, based on investigations of highway landscaping, the Ministry of Highway Engineering of the former Soviet Union developed and issued *Instructions on Highway Architectural Art and Landscape Design*. Landscape design has now become increasingly important in terms of highway design, and many countries have accordingly revised and issued relevant specifications and regulations.

European countries feature high forest coverage, good vegetation protection and appealing motorway landscaping. For the construction of the motorways, close attention has been paid to the protection of trees and restoration of vegetation on both sides of the motorway. For instance, the two directions of many motorways are separated so as to avoid damage to the hilltop, villages, woods, and rivers along the route (Figures 1 and 2). A service area near Stratford-upon-Avon in the UK is built on a hillside in order to preserve the

Figure 1   A motorway detours around a hill to protect the forest.

Figure 2   The two directions of the motorway are not at the same level in order to adapt to terrain features.

existing landscape, and the parking area, petrol station, stores and restaurants are all constructed according to the terrain features, surrounded by the original trees and grass on the hillside. On the motorway from London to Edinburgh, trees were planted at the entrance to a tunnel and at the top of the tunnel upon completion, in order to fit with the surrounding natural vegetation and woods and create an illusion that the motorway is running through woodland and grassland. In the light of local conditions, greening of the central reservation is achieved by accentuating natural and diversified features without pruning weeds, therefore the motorway can be integrated into the natural landscape on both sides.

France attaches great importance to culture and history. As for leisure facilities along motorways, they integrate natural and cultural elements and pay special attention to local customs and practices as well as local landscape features. For instance, trees and flowers are planted in service areas, including tranquil tree-lined trails with an enchanting view; durable and solid facilities for relaxation and recreation such as tables and chairs also fit into the surrounding environment; most of the car parks are in an area covered by green shade, thus creating space for both parking and rest. In addition, travellers have the opportunity to communicate with local residents to understand local life (Figure 3, Figure 4 and Figure 5). In Germany there is a large amount of forest, and many motorways pass through forest for a long distance. Numerous signs are provided on both sides of the motorway to remind drivers of passing wildlife such as red deer and foxes. Rest places for travellers are mostly built in forests along the route, thereby integrating the motorways with the natural surroundings (Figure 6). Motorway design in Germany is characterized by smooth, rational and natural design and rarely involves large-scale cutting and filling. Built based on terrain features, these routes seamlessly blend

Figure 3   A service area in a suburb of Paris, France.

Figure 4   A parking area on a motorway in a suburb of Paris, France seems to be set in a forest.

into the natural scenery, crossing over valleys, mountains and hills along the way. Viewed from a distance, the motorways in this country look like ribbons winding through the green mountains and rivers. To avoid damage to the natural ecology of hilly sections, opposing lanes of traffic are separated; sections with picturesque scenery are generally left in their original state; planting and greening are generally carried out for filled hills, improving the landscape and reducing noise. Excavated sections generally have gentle side slopes with various local plants growing on the slope; masonry work is seldom performed to protect the subgrade.

Figure 5   Picnic benches for resting travellers in a service area on a motorway in Paris, France.

Figure 6   Motorway landscape in Germany.

Motorway landscaping in, for example, Italy, Switzerland and Austria is also designed to be visually appealing, as shown in Figure 7, Figure 8, Figure 9 and Figure 10.

Compared with the good motorway network in European countries and the USA, Asian countries have a relatively weaker network. However, Asia has experienced rapid development in recent years. Despite a land area of only just over 370 000 square kilometres, Japan boasts a modernized motorway network with a total length of 11 520 km. Based on its development plan, Japan will expand its motorway network to a total length of 14 000 km by 2015.

Figure 7   View from either side of a motorway in a suburb of Venice, Italy.

Figure 8   Scenery along a motorway in Zurich, Switzerland.

Figure 9   Corner of a parking area on a motorway near Vienna, Austria.

Figure 10   Recreational area on a motorway near Salzburg, Austria.

Japan's motorways are designed in a rational manner concordant with the terrain features. Motorways are integrated with nature, reducing large-scale cutting and filling. Sometimes the two directions of the motorway are separated at varying heights, reducing excavation work without causing great damage or impact on the surroundings. In addition, Japan puts much emphasis on vegetation protection and restoration as well as natural landscaping in motorway construction. For example, biological protection techniques are adopted for filled, sloped and truncated sections. Trees, shrubs and herbs are all arranged in a natural and varied manner, and rubble masonry is seldom used. In recent years, Japan has adopted the following principles for the development of its motorway network: landscaping is incorporated into motorway construction and an appealing and convenient motorway network that is harmonious with nature along the route has been quickly established in order to satisfy people's demands for a better quality of life. Thus we can see that Japan takes all factors of landscaping, ecology and environment into consideration simultaneously while designing and constructing motorways (Figure 11, Figure 12 and Figure 13).

Since the first motorway (the Shanghai–Jiading Motorway) was completed and put into operation in 1988, the Chinese mainland has also built the Shenyang–Dalian Motorway and the Beijing–Tianjin–Tanggu Motorway. In the space of less than 20 years, the total length of motorways in this country has increased from 18.4 in 1988 to 46 200 km in 2007, ranking second in the world. In accordance with the national motorway development plan, China will build a motorway network with a total length of 85 000 km by 2020.

After nearly 20 years of motorway development, China has begun to focus on motorway ecology and landscaping in addition to engineering standards. China specifies the requirements for ecological protection and environmental construction, requiring an end to irregular excavation, random construction

Figure 11   Motorway landscape in Japan.

Figure 12   A service area of a motorway in Japan built on a hilltop.

and behaviour that destroys the ecological environment. Motorway construction is subject to the requirements of 'ecological and environmental protection and landscaping'. In terms of afforestation, according to relevant authorities, green belts on both sides of motorways will be designed to protect the subgrade and side slopes, reduce soil erosion, restore the ecological environment and enrich the motorway landscape; greening of central reservations is designed to prevent glare, ensure safety and maintain a visually appealing view; the layout of interchanges focuses on using plants for landscaping, and environmental greening and landscaping work in service areas shall be strengthened to provide a comfortable environment for drivers and passengers. A scenic environment in harmony with nature can therefore be realized through landscape design.

Figure 13   Motorway landscape in Japan with a parking area built on the slope.

Jiangsu Province is witnessing rapid motorway development at a rate of over 300 km per year and was the initiator of motorway landscaping in China. In 1999, Jiangsu issued *Guidelines for Jiangsu Motorway Greening Plan and Design* and *Technical Specification for Greening Construction*, so as to standardize motorway landscape design in the province and promote natural and ecological motorway design. For example, the Fenshui–Guanyun Motorway, completed in 2002, focuses on topographic renovation and is a perfect example of a motorway that has taken ecological factors into consideration. The Nanjing–Hangzhou Motorway, built in two phases, respectively, in 2003 and 2004, provides a high-speed connection between Nanjing, the capital of Jiangsu Province, and Hangzhou, the capital of Zhejiang Province. The motorway runs through a part of China reputed to be 'Paradise on the Earth', with a moderate climate, abundant natural resources, a splendid natural environment, strong cultural and historical heritage and a vast wealth of tourism resources. In the wake of further reform and opening up in China, this region has seen remarkable economic growth and plays a leading role in many sectors. The Party Committee, the provincial government and the provincial Ministry of Transportation required that the motorway should be built as the first 'green motorway' in Jiangsu integrating both ecological and environmental protection, landscaping and tourism.

To achieve this goal for the Nanjing–Hangzhou Motorway, the Jiangsu Motorway Construction Headquarters introduced the advanced motorway design concepts and methodologies from foreign countries to conduct landscape design according to international standards. With the introduction of the state-of-the-art 'bead chain' design concept and through elaborate design and construction, the completed Phase I of the project represents a well landscaped motorway integrated with the surrounding natural environment. Damage to the environment along the motorway was minimized in construction, and therefore this noticeably improved the appearance of the surroundings.

The construction of the Nanjing–Hangzhou Motorway has significantly improved motorway construction in Jiangsu, and has played a guiding and demonstrative role for motorway landscape construction within and outside the province. It is important to study and summarize the advanced design concepts and adopt the advanced technologies and methods, so as to promote fully motorway landscape construction in Jiangsu Province.

1 Impact of Motorway Construction on the Environmental Landscape and Protection Countermeasures

1.1 Impact on the natural environment

Topography

Motorways are large-scale man-made projects. During their construction, the topography of the surrounding area will be subject to significant impact. For example, before the construction of the Nanjing-Hangzhou Motorway, the area was characterized by beautiful scenery (Figure 1.1). However, since its construction, great changes have happened to the local environment. First, the motorway has occupied a large area of arable land. During Phase I of the project alone, 4329.86 mu (about 288.66 hm^2) of land and 2205.3 mu (about 147 hm^2) of excavated land was requisitioned; there was a fill volume of 6.1732 million m^3 and an excavation volume of 3.1279 million m^3. As such a large area of cultivated land was requisitioned, the topography across the construction area of the surrounding area was subject to serious degradation (Figure 1.2), especially the areas of great natural beauty. For example, over 400 000 m^3 of stone was excavated from the Donglushan Mountain in Lishui County, resulting in an 80 m high rocky slope surface (Figure 1.3). The construction of the motorway therefore resulted in heavy damage to the local environment. Of course, we can mitigate damage to the local environment by planting vegetation and covering the rock surface with plants, but it is difficult for the environment to revert back to its original condition and therefore destruction to the environment is irreversible. Data show that the area needed to construct a motorway is larger than an ordinary road or railway. In flat and hilly areas, the land occupation ratio of a motorway is usually 8.0~10.7 hm^2/km. Table 1.1 shows the land occupation of the Nanjing-Hangzhou Motorway (Phase I).

The Environment and Landscape in Motorway Design, First Edition.
Qian Guochao, Tang Shuyu, Zhao Min and Jing Chun.
© 2014 China Communications Press. Published 2014 by John Wiley & Sons, Ltd.

Figure 1.1   Picturesque scenery before construction of the Nanjing-Hangzhou Motorway.

Figure 1.2   Excavation resulting in destruction of the landscape.

Figure 1.3   Excavation resulting in a steep slope on the Donglushan Mountain.

Table 1.1   Land occupation area of the Nanjing-Hangzhou Motorway (Phase I) (unit: mu).

Points of origin and destination	Township	Farmland	Commercial forest	Hilly land
K40+480~K50+480	Zaicheng	1063.7	118.2	23.6
K50+480~K52+300	Donglu	0	118.7	79.1
K52+300~K59+156.1	Baima	661.1	103.2	11.5
K59+156.1~K60+577.77	Gonghe	35.3	0	113.4
K60+577.77~K61+380	Shangxing	15.8	0	66.8
K61+380~K68+399.41	Jiuxian	474.4	412.4	45.8
K83+000~K99+750	Xinchang	1271.9	300.5	137
K99+750~K103+706.66	Chating	289.7	0	31.8
K103+706.66~K111+538.29	Chengnan	593	45.6	19.6
K111+538.29~K112+121.97	Yantou	42.7	3.3	1.4
K112+121.97~K116+000	Jingtang	691.3	317.7	53.0
K116+000~K122+126.5	Guijing	190.2	380.3	63.4
K122+126.54~K130+781.97	Huankeyuan	413	239.4	148.1
K130+781.97~K146+960	Dingshu	673.9	868.4	293.4
K146+960~K149+462.23	Fudong	77.7	136.4	45.3
Total		7374.7	3265.8	1168.6

1 mu is equal to ~666.67 m².

According to Table 1.1, the Nanjing-Hangzhou Motorway occupies an area of 11 808.4 mu (about 787.23 hm²), most of which is farmland and commercial forest, which amounts to 10 640.5 mu (about 709.37 hm²). As a consequence, it brought about not only a devastating effect on the local natural environment, but also had an adverse impact on local agricultural production and socio-economic development.

Soil erosion

Soil is one of the most important factors for the growth of vegetation along a motorway. During motorway construction, quarrying, borrowing earth and spoil grounds will cause soil erosion. The fertility of soil is reduced, changing its physicochemical properties, which makes it more difficult for vegetation to grow and recover. If the soil structure of the land under construction changes due to compaction from machinery or the land being trampled on, the fertility of the soil may not recover for a long time.

Soil erosion destroys the vegetation along a motorway. In turn, the loss of vegetation further exacerbates soil erosion. This vicious circle makes it very difficult for the vegetation to grow and recover. On the other hand, soil erosion causes change in topography, resulting in water and soil loss (Figure 1.4). According to a survey, water and soil loss resulting from motorway construction in Guangdong Province reaches 4.6203 million m³ every year. For example, during the construction of the Ningxia Guyaozi – Wangquanliang Motorway, the area affected by water and soil loss increased by 592.48 hm². This was due to disturbing the original topography and damage to the soil

Figure 1.4   Soil erosion.

Table 1.2   Predicted value of the potential intensity of soil erosion on the Nanjing-Hangzhou Motorway (Phase I).

Section	Topography	Current situation of land		Construction period			
		Erosion modulus (t/km^2 yr)	Grade	Subgrade slope (t/km^2 yr)	Grade	Borrow area (t/km^2 yr)	Grade
K40+480~K67+920	Hilly land	1500	II	6225	IV	5505	IV
K67+920~K83+000	Flat to undulating hills	1500	II	6225	IV	5505	IV
K83+000~K95+840	Low mountains and hills	3000	III	12450	V	11010	V
K95+840~K115 +205	Flat to undulating hills	1500	II	6225	IV	5505	IV
K115+205~K133 +61	Low mountains and hills	3000	III	12450	V	11010	V
K133+612.5~K148 +127	Flat to undulating hills	1500	II	6225	IV	5505	IV

and vegetation. Of the 592.48 hm^2 affected, cultivated land amounted to 134.20 hm^2, woodland 152.10 hm^2, grassland 295.75 hm^2, and other land 10.43 hm^2. For soil erosion on the Nanjing-Hangzhou Motorway, see Table 1.2.

According to Table 1.2, the construction of the Nanjing-Hangzhou Motorway has resulted in serious soil erosion.

Climate

Upon construction, motorways can generate a microclimate environment, which mainly depends on the properties of the underlying surface and the composition of the atmosphere. A motorway microclimate that is adverse to plant growth has the following features:

(1)   Cold. Motorways are generally far away from centres of population, with high subgrades, spacious topography and fast air convection. Particularly on interchange sections, the clearance under the bridge is relatively high, resulting in freeze injury to trees on windward slopes in North China. In the winter of 1997, one-third of Chinese Juniper in the central reservation died or was affected by frostbite due to the wind beside the approach of Shahe Grand Bridge 233 km along the Beijing–Shijiazhuang Motorway.

(2)   High temperature. At either side of the central reservation, there is a 10-m-wide black paved surface. During the summer, the air temperature can reach roughly 40°C, but after factoring in the thermal radiation of the black pavement, the local air temperature may reach over 50°C in some areas.

(3)   Drought. Due to the thermal radiation of the black pavement, the high temperature exacerbates the evaporation of soil moisture and the transpiration of trees, resulting in the death of trees due to drought. This phenomenon is even more apparent in the central reservation.

(4)   Strong gale. Motorways feature high subgrades, spacious topography and fast convection. High-speed vehicles increase the wind speed, especially large buses, which may produce an instantaneous wind speed of 25 m/s. This is a very common occurrence, which causes the trees to sway wildly, resulting in damage to their root systems and exacerbation of water loss and drought.

Hydrology

In motorway construction the direction of surface water may often change its course. Due to the diversion of the river flow, water and soil loss worsens at areas where the flow of water is concentrated and erosion occurs where the structure of the water flow is adverse. In river or wetland areas, it is necessary to change the original direction of the river when a motorway is built. After the direction of the river has been changed, a very large flow is generated in areas where many waterways come together and the flow rate speeds up. The hydrological conditions then change, resulting in floods, a worsening of water and soil loss or an increase of downstream silt (Figures 1.5 and 1.6). The area where the Nanjing-Hangzhou Motorway (Phase I) was built is in the Qinhuai River basin. Along the route, there are low mountains, hills and flat areas as well as numerous big and small lakes and crisscrossing rivers. Most of the waterways that the Nanjing-Hangzhou Motorway passes over are barge routes and irrigation channels. Although the aim was to minimize and avoid adverse effects during bridgework, culvert design and subgrade construction, elements

Figure 1.5   The motorway results in the redirection of the original waterway.

Figure 1.6   Water and soil loss due to the destruction of vegetation.

including the navigation, irrigation and drainage of the existing waterways were still damaged or impaired.

Vegetation

The impact of a motorway on vegetation mainly refers to the direct impact of land destruction, borrowing earth and spoil grounds during motorway construction, as well as the indirect impact of motorway traffic. The former is transient and irreversible; the latter is more long-term and reversible. During motorway construction, direct destruction of vegetation mainly stems from the following two aspects: permanent destruction from site clearing, and damage from the temporary spoil ground and construction road. For example, the land permanently occupied by the Jilin–Changchun Motorway project is 571.8 hm², of which dry farmland accounts for 76.5%, paddy fields 12%, uncultivated land 1.9% and woodland 9.6% (19 360 timber trees were felled). According to calculations, the amount of carbon dioxide absorbed annually by the vegetation has decreased by 8274 t and the amount of oxygen released annually has decreased by 6040 t. Due to the occupation of cultivated land, the amount of oxygen released annually has decreased by 1500 – 5000 t. Using Zhoukou – Shengjie (Provincial Boundary) Motorway in Henan as an example, the 9.498-km-long section within Beijiao Township, Shenqiu County alone covers a cultivated land area of 69.8 hm^2, equal to a vegetation damage rate of 1.968%. As a large amount of cultivated land has been occupied, farmers may cut down woodland to make up for the occupied land, further worsening the damage to the vegetation along the motorway (Figure 1.7). After the completed motorway was opened to traffic, pollution from vehicles has also been an indirect factor contributing to vegetation damage. It is reported that farmland within a range of 50 m of both sides of the motorway has seen an average reduction in output of 15% due to dust and exhaust fumes from trucks transporting coal from Shanxi Province to other regions. As a result, the wheat yield in the province has decreased by 28.13 million kg every year, equal to a loss of RMB 33.75 million.

Of the land requisitioned for the Nanjing-Hangzhou Motorway (Phase I), arable land amounts to 150.02 hm^2, commercial forest 50.17 hm^2, hilly land 22.68 hm^2, borrowed land 147.02 hm^2, as well as extra temporary land use for the construction road, mixing yard, prefabrication yard and construction camp. The vegetation at these places was cleared during construction. After completion of the project, new vegetation was planted over an area of 909 029 m^2, accounting for 20.76% (6535.16 mu or ∼436 hm^2) of the total land requisitioned. In other words, the Nanjing-Hangzhou Motorway (Phase I) resulted in a loss of green vegetation of 271.90 hm^2 (Figure 1.8) to this area; it was therefore inevitable that the local environment would be affected.

Figure 1.7   The motorway occupies the local tea plantations and woods.

Figure 1.8   Forest vegetation in the Donglushan Mountain.

Wildlife

Wild animals are the main victims of habitat fragmentation during motorway construction. As the forest decreases in size and is divided by residential areas and traffic networks, their habitat gradually shrinks. These small habitats are not sufficient for the animals to survive. If their territory is too small, they cannot acquire enough food, resulting in the decline or extinction of these animal populations. Increased traffic flow and expanding human activity also reduces their habitats, affecting their mating and reproduction, further aggravating the impact on them. In addition, newly built motorways directly cause a loss of habitat and terrain features, resulting in a change in climatic factors such as sunlight, wind speed, temperature and humidity. Meanwhile, vibration, noise, atmospheric pollution and soil pollution from vehicles has a negative impact on the survival, reproduction and migration of local plants and animals.

According to a survey, there are fewer wild animals and species in areas next to the Nanjing-Hangzhou Motorway (Phase I), including species such as wolf, boar, vole, hedgehog, yellow weasel, rabbit and common birds. Although the construction of the motorway has resulted in adverse impacts on such wildlife, the consequences are not too severe as there are no rare or state-protected species in the region.

1.2   Impact on history, culture and scenery

If the route is not selected carefully or if no attention is paid to conserving water and soil during construction, there may be the following problems:

(1)   The motorway passes through a town.
(2)   It passes through a scenic spot.

(3)   It goes through a forest.
(4)   Mountains have to be cut into and deep holes must be filled.
(5)   The flow of the river changes and streams become blocked.

Poor route selections such as these not only damage the ecological environment across the region but also the integrity of the history, culture and scenery in the area, as well as the regional tourism resources. A survey has found that there are no cultural relics and scenic spots requiring special protection along the Nanjing-Hangzhou Motorway (Phase I), so the impact on the history, culture and scenery of the area along the motorway has been small and the prospect for the development of tourism resources is promising. Phase II passes by the Longbeishan Mountain National Forest Park but does not go directly through it, so its impact is negligible. The completed Nanjing-Hangzhou Motorway serves as a protective barrier for the Longbeishan Mountain National Forest Park, protecting it against the erosion of land for urban development.

1.3   Pollution of the surrounding environment

Motorway construction has polluted regions along the route to a different degree.

Noise pollution

There are two kinds of noise pollution: one is from excavators, bulldozers, land levellers, lorry mounted mixers and other types of construction vehicles. These vehicles and equipment create significant noise pollution. For example, the noise from a common road building machine exceeds 80 dB, while that from a pile driver is higher than 100 dB. The other kind of noise pollution is from traffic, forming a noise belt along the motorway route. The noise from these sources has a negative psychological and physiological impact on constructors and people in the surrounding areas. It also reduces people's work efficiency, having a particularly apparent impact on sensitive areas with a high population density on both sides of the motorway (schools, residential areas, commercial districts, hospitals, etc.). At nighttime, the noise affects the lives of people living alongside the motorway and the impact will only worsen in the future (Figure 1.9).

Water pollution

There are two sources of water pollution from the motorway: one is from subsidiary facilities along the route, such as service areas, toll stations, management centres and maintenance work zones. In general, a motorway can produce about 200 000 t of sewage a year. After treatment, the sewage can

Figure 1.9   Residents on both sides of the motorway will be disturbed by noise.

Table 1.3   Predicted value of the concentration of pollutants in the surface runoff.

Item	SS	BOD5	Petroleum	CODcr
120 min mean value (mg/l)	100	5.08	11.25	107
Quantity of pollutants discharged (t/a)	2977.8	15.1	33.5	318.6

meet effluent standards, that is CODcr 100 mg/l, BOD 530 mg/l, SS 70 mg/l and petroleum 10 mg/l. However, this waste water will result in water pollution if discharged into a river or a lake. If the waste water is not treated or if treatment is not complete or does not meet the necessary standards, the resulting water pollution is more severe and harmful.

The other source of water pollution is pollutants from automobile exhaust fumes, surface residues and surface materials. These pollutants flow across the surface of the road into reservoirs and rivers after rain, contaminating these areas to a certain degree. In the region where the Nanjing-Hangzhou Motorway (Phase I) is located, the mean annual precipitation is 1050 mm, and the total area of pavement is 2.836 km², therefore the annual surface runoff volume is 2 977 800 m³; the total discharged quantities of all types of pollutants are shown in Table 1.3.

Atmospheric pollution

The main cause of air pollution is automobile exhaust emissions, including carbon monoxide (CO), nitric oxide (NOx), total hydrocarbons (THC) and total suspended particulates (TSP), of which NOx is the most serious pollution to the environment, followed by CO and THC.

After the Nanjing-Hangzhou Motorway was completed there was an increase in NOx emissions because vehicles started to travel much faster. The motorway does not pass through urban areas, thereby reducing congestion on urban roads and relieving air pollution of cities along the motorway.

Figure 1.10   The storage yard is a source of on-site dust pollution.

Dust and waste

Dust from construction sites, storage yards, uncompleted pavement and construction side roads results in serious pollution of the surrounding environment. Relevant testing shows that the concentration of TSP is 5.097 mg/m^3 150 m downwind of a construction vehicle, higher than the national ambient air quality Grade II. As the wind speed increases, the area affected by dust becomes larger (Figure 1.10).

The household refuse and construction waste produced by construction workers has a negative impact on the surrounding environment, such as visual pollution and damage to the beautiful surroundings and vegetation. Examples include large quantities of building materials and other materials abandoned during transportation, stones and earth left behind after demolition work, household refuse and construction waste from construction camps, refuse from service areas, parking areas and management centres as well as litter dropped by drivers and passengers. This litter is mainly waste paper, plastic bags, plastic boxes and plastic bottles. As the number of vehicles increases in the future, pollution from litter will become more severe.

1.4   Impact on public activities and production

Because the Nanjing-Hangzhou Motorway is totally enclosed with a complete interchange system, it separates villages near the route. The Nanjing-Hangzhou Motorway (Phase I) passes directly through 5 villages and there are 9 villages about 15–100 m away from its main central line. The separation that the motorway causes results in inconvenience and difficulties for farmers working in the fields and moving from one area to another, as well affecting administrative divisions and regional planning (Figure 1.11). After land has been occupied by the motorway, some farmers lose their land

Figure 1.11   The motorway isolates the village on both sides.

and therefore have less income, resulting in a loss to the regional economy. Meanwhile, the resettlement and relocation of residents due to the project seriously affects farmers' everyday lives and is an added psychological burden.

1.5   Landscape protection countermeasures and measures

In order to prevent and alleviate the damage and adverse impact on the surrounding region during construction, it is important to focus on two stages of the project: planning and operating. Various measures should be taken to solve problems in design, construction and operation. Special countermeasures should be introduced for environmental protection, foreseeing hidden dangers and taking measures to eliminate them. All measures should be adopted in order to solve unavoidable problems that damage the environment. With regard to adverse impacts that have already occurred, measures should be taken to remedy and correct them. There are four specific measures that can be taken.

Select a rational route and minimize damage to the environment

Use new technologies to optimize the design

Selecting and determining a good motorway route design is a prerequisite for building a good motorway. First, it is necessary to investigate and survey the natural and social environment of the regions concerned, and carry out a detailed analysis on the data in order to determine an optimal plan. 3S technology for survey and design can not only enhance survey precision, but also can lower the survey cost, ensuring the most optimal motorway route selection. 3S technology refers to the integration of remote sensing technology (RS), the geographic information system (GIS) and the global positioning system (GPS), which can better simulate the hydrological, geographical, geological and climatic conditions of the regions where the motorway would pass through, thereby selecting an optimal motorway route.

Bypassing urban areas to coordinate with urban planning

When selecting a motorway route, it is best to bypass towns, natural conservation areas, water conservation districts, scenic spots, historical sites and tourist attractions, in order to prevent or mitigate the motorway's impact on the environment (Figure 1.12). The route selected should avoid interfering with the development of towns and help drive their further economic development. For this reason the distance between the motorways and towns should be chosen appropriately. The distance between the Nanjing-Hangzhou Motorway and Lishui, Liyang and Yixing is about 1, 1.5 and 2 km, respectively, therefore avoiding interfering with the urban planning of these three regions. Meanwhile, the route stays away from large-scale infrastructure and agricultural centres of excellence located in some cities.

Reduce land occupation and protect agriculture

Land occupied for motorway construction is permanent, therefore, the occupation of arable land should be minimized during planning, especially high-yield plots and commercial crop areas. More earth should be borrowed from barren hills in order to save arable land. As shown in Figure 1.13, no borrow pit is excavated next to the Nanjing-Hangzhou Motorway in the Taihu Lake region in order to protect precious farmland resources there. In order to save land, the central reservation of the 2 km connecting section between the Nanjing–Gaochun Motorway and the Nanjing-Hangzhou Motorway is of narrow design, being only 2 m wide.

Figure 1.12   The motorway keeps away from the ancient pagoda in Yixing and passes by Longbeishan Mountain National Forest Park.

Figure 1.13   The central reservation between Luojiabian and the toll station on the main lane is 2 m wide.

Reduce the amount of excavation and fill to conserve natural vegetation

The motorway alignment should integrate fully with the local topography. As long as the requirements for motorway standards are met, the alignment should adapt to topographical changes as far as possible, integrate with the natural environment and ensure the reduction of the amount of fill and excavation in order to alleviate damage to the original land vegetation. For example, a tunnel was built through the Tizishan Mountain so as to reduce the impact of the Nanjing-Hangzhou Motorway on the natural mountains in the Taihu Lake scenic area, thereby conserving the natural environment (Figure 1.14). Shilibei Village, 5 km north of Lishui County town, Nanjing, intersects with the Changzhou–Liyang Motorway, where there is an ancient Ginkgo bilobamaidenhair tree within the planning boundary line. According to legend, this ancient tree was planted in the Ming Dynasty (1368-1644 AD).

Figure 1.14   The tunnel in Tizishan reduces damage to the natural environment.

Figure 1.15   Motorway geometric design is adapted to topographical changes.

In order to protect it, the villagers were willing to dismantle more houses. After repeated research, the construction headquarters decided to alter the design scheme – changing the cut slope into a vertical type retaining wall. In other countries, people pay more attention to the adaption of the motorway to the local topography when selecting motorway alignment. Sometimes the two sides of the motorway are not at the same height (Figure 1.15), thereby reducing the excavation of earth and rock and minimizing the impact on the surrounding environment. Motorways should be integrated into the natural and social environment as far as possible.

Design passageways to resolve isolation issues

Motorways are in general enclosed networks, and this will unavoidably result in inconvenience to the lives and agricultural production of residents nearby, as well as isolating the habitats of wildlife populations. For this reason all kinds of passageways should be considered during design. The Nanjing-Hangzhou Motorway (Phase I) is constructed with 54 passageways, 102 culverts and 8 overpasses for tractors and people, which essentially meet the production and daily needs of people living along the route as well as ensuring sufficient habitat for wild animals (Figure 1.16).

Attach importance to environmental protection projects during construction, alleviate pollution and prevent water and soil loss

Rational use of land resources

During motorway construction, the construction procedures should be determined first. Temporary land occupation should be minimized, the occupation time should be shortened and the land should promptly be restored to its

(a)

(b)

(c)

Figure 1.16   (a) Overpass, (b) passageway and (c) culvert.

original function. With regard to temporary land use for stock grounds and mixing yards, land within the scope of the motorway should be selected, such as interchanges, service areas and toll stations, so as to ensure that additional land is not occupied. If conditions permit, the bridge construction site may be arranged at the intersection or on the banks of the river. Temporarily occupied farmland should revert back to farmland immediately after construction.

Prevent water and soil loss

During motorway construction, a huge amount of excavation take place, such as subgrade, retaining wall and slope protection works. With regard to the area where there may be direct surface runoff during rain, the sedimentation basin and geotextile fence should be constructed in advance to prevent water and soil loss. Thus in the case that muddy water flows through the sedimentation basin slowly, silt will be deposited. All the engineering works for soil conservation, soil consolidation and preventing water and soil loss throughout the motorway should be carried out simultaneously with the main works and subject to handover and acceptance. The Nanjing-Hangzhou Motorway project was an innovative model in slope protection technology. The gradient of the side slope is reduced as far as possible, to a maximum of over 1:6. The hard side slope constructed with mortar rubble was changed to a soft soil slope for ecological protection, thereby expanding the landscaping and vegetation area. As such, spoil and rock were used as subgrade fill. The rock disposal dump site was used for landscaping. For example, a sightseeing pavilion was built at the rock disposal dump in Donglushan after being refilled with earth, greatly improving the surrounding environment (Figures 1.17 and 1.18). During construction along the waterways or when constructing bridge foundations, spoil and other waste are not allowed to be dumped into rivers, lakes, reservoirs or irrigation canals in order to protect water sources and waterways. In addition,

Figure 1.17   The stone is filled with earth for greening.

Figure 1.18   A pile of rocks becomes a scenic spot.

the subgrade and the surface drainage system should not destroy the original river system and surface runoff mechanism.

Prevent and control pollution

During construction the main sources of environmental pollution include noise and dust. The noise stems from construction machinery and transportation vehicles. Noise should be strictly controlled in accordance with the *Noise Standards on Both Sides of Arterial Traffic*, and *Noise Limits for Construction Sites*, to reduce noise as far as possible. Measures generally taken to reduce noise are to emphasize safe and civilized production and strengthen the management of construction machinery and transport vehicles to ensure that noise from construction machinery conforms to the limits. Other countermeasures include adjusting construction and operation times, avoiding construction during the night, concentrating the operation of noisy machinery within a certain time, completing construction as soon as possible, preventing noise pollution from blasting, adjusting the quantity of construction machinery in operation and compensating those who suffer from noise.

Dust causes air pollution during construction. A wind shield wall (net) should therefore be erected around the material store yard, and especially around the fly ash pile. The road to the store yard should be sprayed with water frequently; the bituminous mixture should be batched and mixed in a fully enclosed yard or station. The subgrade, should be compacted layer by layer, reducing dust by spraying with water and therefore ensuring that the water quality on both sides of the road is not affected by the dust and rainfall runoff. The bituminous mixing yard should be located downwind of the prevailing wind direction and at least 300 m away from any villages, towns or residential areas in order to prevent pollution from bituminous fumes.

Figure 1.19   The ground is afforested immediately.

Promptly and comprehensively plant vegetation to prevent water and soil loss and pollution

When construction is complete, clear up and level the ground immediately, then plant nursery-grown plants and grass according to the design requirements. These are effective measures to prevent water and soil loss, surface water pollution and the inhalation of harmful gases as well as to prevent dust and reduce noise (Figure 1.19). Land vegetation helps to mitigate the flow of rainwater to the surface soil and slow down the runoff, as well as enhance sedimentation efficiency, filtrate suspended solids, and improve soil permeability. As a result, soil erosion resulting from surface runoff is alleviated and runoff pollution can be checked effectively. As plants develop roots and continue to grow, their ability to conserve water and prevent and control pollution increases (Figure 1.20).

Strengthen environmental quality management during the operation period

Refuse and water from service areas should be treated with suitable waste and sewage treatment equipment based on the amount discharged before being discharged according to required standards. This prevents waterways from being polluted. For example, the sewage treatment facilities in the Tianmu Lake service area uses an advanced recycled water treatment system, and the discharged water meets the grade one standard; if it is filtered and disinfected further, it may be used for purposes such as watering flowers, washing cars and flushing toilets. In order to prevent noise from affecting residents along the motorway, measures against noise should be employed, such as

Figure 1.20   The scenery of the Donglushan Mountain with grass and trees after destruction.

providing a forest belt between the village and the motorway, heightening the fence, installing anti-noise windows and erecting noise barriers. Although the soundproof effect of a forest belt is not as good as other engineering measures, and despite that fact that a large area of land is required, forest belts can afforest and improve the surrounding environment in addition to reducing noise, and conserving water and soil. They help to restore the local ecological environment and make the environment more pleasant for people to live in. Forest belts are therefore often employed as a preferred way to reduce noise.

In the period of motorway operation, air quality on both sides of the route declines due to pollution from automobile exhaust fumes. The main pollutants include CO, NO_2 and THC. By monitoring the air quality in these regions, it has been observed that among the pollutants, 65–80% of CO, 50–60% of NO_2 and 80–90% of THC come from automobile exhaust emissions. It is therefore crucial to improve air quality along the motorway by reducing emissions from automobile exhausts. At present, the most effective ways to reduce emissions from automobile exhausts include improving operation and engine performance, modifying the fuel used and treating automobile exhaust fumes.

Restore and create natural landscapes by planting vegetation, enabling the motorway to integrate with the surrounding environment and promote the development of tourism

Great efforts should be made to design the landscape of the whole motorway project. By planting vegetation to improve the landscape, the construction of the motorway will ensure the integration of local landscape, environmental protection and tourism. A good environmental design not only improves the natural environment along the route, it also enriches the landscape and fully maximizes the potential for tourism in the region, while eliminating or alleviating driver fatigue and greatly reducing traffic accidents. The following actions should therefore be taken.

Plant vegetation in all the exposed areas and restore the surrounding ecological environment to its original state

All the exposed areas within the separation fence should be planted with trees, flowers and grass, especially with native tree and grass species. The structural features of the natural plant community should be simulated as far as possible to eliminate the adverse impact of construction on vegetation and the environment along the motorway as well as to lessen the degree of damage. For this purpose, within the 31-km-full length of the Nanjing-Hangzhou Motorway (Phase I), the afforested area amounts to more than 900 000 m^2. Large scale greening of the road area is also a positive measure to conserve water and soil and improve the surrounding environment.

Build themed landscapes

Motorway landscapes should be both integrated into the natural environment and visually appealing. For example, many attractive natural landscapes have been created within the interchange area and sensitive zones along the motorway. By planting large areas of tall trees and shrubs, redirecting water and rearranging rocks, many landscapes such as woodlands, grasslands, fields and wetlands have been created. These man-made landscapes fit seamlessly into the natural environment and highlight local cultural characteristics, forming a more attractive landscape for visitors.

Use ecological protection technology to cover rocky slopes with vegetation and restore the damaged ecological environment to its original state

With respect to large areas of cut rock and side slopes, a series of new technologies and methods should be employed for restoration as it is difficult to restore vegetation. Examples of this new technology include 'soil replacement and sowing', 'soil consolidation with straw bags' and 'soil consolidation with straw sticks'. Through such new technologies and methods, green vegetation covers the rocky slopes again, which successfully blends into the surrounding vegetation, thereby improving and enhancing the quality of the local environment and landscape.

2 Environmental Landscape Design of Filled and Excavated Sections

2.1 Concept

During subgrade construction the terrain and topography along the route should be renovated according to the design requirements for subgrade elevation. Sections higher than the subgrade elevation should be 'excavated', while sections lower than the subgrade elevation should be 'filled'. The subgrade slope is a slope resulting from excavation, called the upper slope or cut slope, also known as the excavated section. The slope that results from filling is called the lower slope, also known as the embankment slope or the filled section. During construction, the original vegetation and soil are destroyed regardless of whether the slope is formed from filling or excavation. Not only does the exposed slope affect the landscape, but it is also very prone to erosion by rain, resulting in water and soil loss, or landslides (Figures 2.1 and 2.2). In order to prevent the collapse of steep slopes, engineering protection measures are taken, such as building the retaining walls in stages and using mortar rubble or prefabricated cement slabs. This measure is a good way to reinforce the soil and protect the slope, but the protrusive grey concrete does not fit into the surrounding natural environment. Furthermore, it releases a large amount of radiant heat (Figure 2.3). Following a continued increase in motorway construction standards and environmental quality regulations, it is now a prerequisite that motorways are built such that they blend into the surrounding environment. It is therefore important that biological protection measures are taken to protect the slope at the same time as using engineering protection measures. According to the situation on each individual slope, this includes planting various kinds of vegetation, or employing a combination of biological and engineering protection measures. This is an important step which is now increasingly being taken into account during motorway construction.

The Environment and Landscape in Motorway Design, First Edition.
Qian Guochao, Tang Shuyu, Zhao Min and Jing Chun.
© 2014 China Communications Press. Published 2014 by John Wiley & Sons, Ltd.

Figure 2.1   **Exposed slope due to destruction of the vegetation.**

Figure 2.2   **Destroyed topography due to excavation.**

Figure 2.3   **Rocky steep slope in the Donglushan Mountain due to excavation.**

2.2   Classification of the side slope

Side slopes can be classified as natural or artificial. Side slopes created during motorway construction are artificial. Artificial side slopes may be divided into different categories according to their texture, height, length and gradient. See Table 2.1 for the different side slope categories on the Nanjing-Hangzhou Motorway.

There are many rocky side slope sections along the Nanjing-Hangzhou Motorway. One of these is the Donglushan rocky side slope, which is 1460 m long and 76.4m high. After undergoing renovation, its gradient is now 1:1~1:1.25.

2.3   Functions of planting vegetation on the slope

Planting vegetation on the slopes of the motorway is a fundamental step that keeps the side slope and subgrade stable, and ensures the smooth operation of the motorway. It has the following functions.

Table 2.1   Artificial side slope types on the Nanjing-Hangzhou Motorway.

Basis of classification	Name	Brief description
Texture	Rocky slope	Made of rock; the features and nature of the slope vary with the type and structure of rock
	Soil slope	Made of soil; the nature of the slope varies with the type of soil
Slope height	Ultrahigh side slope	The rocky side slope is over 30 m in height, and the soil slope is over 15 m in height
	High side slope	The rocky side slope is 15–30 m in height, and the soil slope is 10–15 m in height
	Medium–high side slope	The rocky side slope is 8–15 m in height, and the soil slope is 5–10 m in height
	Low side slope	The rocky side slope is below 8 m in height, and the soil slope is below 5 m in height
Slope length	Long side slope	The slope length is more than 300 m
	Medium–long side slope	The slope length is 100–300 m
	Short side slope	The slope length is less than 100 m
Gradient	Gentle side slope	The gradient is less than 15°
	Moderate slope	The gradient is 15–30°
	Abrupt slope	The gradient is 30–60°
	Steep slope	The gradient is 0–90°
Stability	Stable slope	Stability conditions are good so no damage will occur
	Unstable slope	Stability conditions are poor; it is subject to partial damage so it must be treated to maintain stability

Conserve water and soil and stabilize the side slope

The runoff caused by rain often results in the erosion of side slope soil, leading to water and soil loss. Dense branches and leaves can effectively alleviate the impact of rain on the surface soil, thus reducing erosion and water and soil loss on the slope. The dense reticular root system strengthens the cohesion of soil mass. Some plants even excrete organic matter, which then reinforces the soil due to the agglomeration of soil particles. As these plants grow, their protective effect becomes increasingly stronger (Figure 2.4). A test was carried out which involved artificial rainfall on side slopes with different levels of vegetation coverage. The gradient of the side slope was 30° and the rainfall intensity was 200 mm/h. The test results proved that soil loss decreases as the amount of vegetation coverage on the side slope increases. Mature grass can play an effective role in conserving water and soil (Table 2.2). In addition, slope protection plants can reduce the moisture of the slope and lower the pore water pressure of soil mass through transpiration, which is conducive to the stability of the side slope.

Figure 2.4   Planting vegetation on the slope conserves soil and reinforces the slope.

Table 2.2   Relationship between grass coverage and soil erosiveness.

Coverage (%)	100	96	68	36
Erosiveness (%)	0	10	38	82

Improve the appearance of the motorway and restore the surrounding ecological landscape to its original state

According to the variety and condition of the side slope, different plant species should be combined according to local conditions, in order to create different landscapes, improve and enrich the appearance of the surrounding environment and therefore restore and improve the environment damaged during construction of the motorway (Figure 2.5). Meanwhile, green plants can also prevent pollution by purifying and improving the air quality.

Lower slope protection costs

Biological slope protection technology can greatly lower slope protection costs. According to calculations, biological protection costs are 1/7 – 1/10th of the earth – rock protection costs. Even if biological protection is combined with engineering protection, the cost is just one-third of employing only engineering protection.

Figures 2.6 – 2.12 are photographs of biological side slope protection measures to improve the appearance of the environment.

Using data provided by Guangdong Hehui Motorway Co., Ltd, a comparison between investments in biological and engineering slope protection is shown in Table 2.3.

Figure 2.5   Planting vegetation on the slope reinforces the soil and protects the slope.

Figure 2.6   Biological slope protection improves the environment.

Figure 2.7   Biological side slope protection measures.

Figure 2.8   Biological side slope protection measures.

Figure 2.9   Biological side slope protection measures.

Figure 2.10   Biological side slope protection measures.

Figure 2.11   Biological side slope protection measures.

Figure 2.12   Biological side slope protection measures.

Table 2.3   Comparison between investments in biological and engineering protection.

Measure	Comprehensive unit price (approx.)	Main applicable scope
Rubble concrete (mortar rubble) retaining wall, about 2 m in thickness	300 yuan/m²	Lack of sufficient conditions for grading of the steep slope
Mortar rubble facing wall, about 0.5–1.5 m in thickness	250 yuan/m²	Soil and rock slopes
Mortar rubble slope protection, about 0.3 m in thickness	250 yuan/m²	Soil slope, closing
Suspended mesh shotcrete, about 0.08 m in thickness	130 yuan/m²	Rocky slope, closing
Pre-stressed anchor cable, about 20 m in length	450 yuan/m	Lack of sufficient conditions for grading of the steep slope, rock breakage
Arched water retaining skeleton, about 0.4 m in depth	38 yuan/m	Protection against erosion
Spreading grass seeds, about 0.06–0.1 m in thickness	80–100 yuan/m²	Soft rocky slope
Soil replacement and spreading, about 0.03–0.1 m in thickness	30–130 yuan/m²	Hard rocky slope
Three-dimensional vegetative net planting	23–28 yuan/m²	Embankment filled with stone and sand
Hydraulic seed spraying	5–10 yuan/m²	Embankment filled with earth

2.4   Habitat conditions on the slope

(1)   The soil on the side slope is barren and dry. On most upper slopes, there is no surface soil. The exposed soil texture is mostly parent material or rock matter, and there is almost no organic matter. Since the lower slope is formed by applying a bulldozer, the soil is compacted, with organic matter accounting for about 1% and a moisture content of approximately 25%, which demonstrates a better quality of the soil generally than the upper slope.

(2)   Poor vegetation recovery. Both the upper and lower slopes are more prone to water and soil loss than flat areas due to the gradient, and thus the seeds of vegetation in the surface soil are prone to being washed away. As the gradient increases, the natural recovery of the vegetation decreases and it becomes increasingly difficult to restore vegetation using artificial breeding methods. Research has shown that in rainy areas of South China the rate of natural plant coverage on the lower slope is 30% over a 7-month growth period, yet on the upper slope it is only 10%, most of which are unremarkable bryophytes.

2.5   Slope landscape design

Slope landscapes should be designed as per the gradient, length, position and soil conditions of the slope, ensuring that the landscape blends into the surrounding natural and social environment. The design principles are as follows:

(1)   Biological protection is combined with engineering protection. Once the motorway has met traffic demands, vegetation should be planted according to the local conditions. Some slope sections have extremely adverse soil quality and lack the basic conditions for plants to grow. For these slopes, engineering protection measures must first be taken to reinforce the soil and stabilize the slope. At the same time, conditions should be created that enable plants to grow. Biological protection measures should be employed to ensure overall protection of the slope and an attractive landscape.

(2)   Based on slope protection, consideration should be given to improving the landscape and improving the appearance of the motorway. In addition to ensuring slope stability, attention should be paid to landscape design and ensuring that the landscape fits into the surrounding natural, social and cultural environment. These resources should be fully utilized to improve the surrounding view.

(3)   Whilst priority should be given to ecological and social benefits, consideration should also be given to economic benefits. When selecting plants, it is important to follow the principle of matching species with the site in question. Based on the results of vegetation studies, there should be

large-scale planting of wild, robust and adaptable native trees, shrubs, vines, bamboo, herbaceous plants and flowers. According to individual habitat conditions on the slope, it is advisable to choose sowing, planting or breeding methods or measures as appropriate. Soil replacement and spreading should be used widely to guarantee biological protection whilst also saving on costs as far as possible.

(4) It is important to follow ecological principles. A plant community that blends into the natural environment should be chosen to protect the slope and rectify environmental damage made during motorway construction. By utilizing plants' evolutionary changes, various kinds of plants with different ecological characteristics should be planted together according to the local conditions. This can ensure the construction of a green and eco-friendly motorway slope protection system with a favourable structure and layout as well as a stable and diverse plant landscape.

(5) A combination of a systematic and diverse approach should be used to accentuate regional characteristics, blend into the surrounding environment and create diverse landscapes. At the same time as ensuring the protection of the slope, once overall unity with the neighbouring environment has been assured, landscapes should be created based on the environmental conditions of different slope surfaces, achieving an effect that is simple yet not monotonous, varied yet not disorderly. The aim is to create an imposing and visually appealing effect by constructing a green and eco-friendly motorway with diverse landscapes.

2.6 Landscape design types

As for biological slope protection and landscape engineering, local climate, soil, vegetation, geology, topography and cultural traditions shall first be investigated, and the design of different sections shall be decided according to the results. Then, the combination of plants shall be designed based on engineering protection measures, slope soil properties, green areas, the direction of traffic and various requirements on features and landscape effects. Due to the diversity of side slope types, the landscape design of motorway side slopes is also varied; the main types are given in the following.

Grass and ground cover plants for slope protection landscaping

This type of landscape design is one of the main slope landscaping forms, made by a particular kind of grass or from grass mixed together with ground cover plants. The lawn consisting only of grass is flat and elegant, and the lawn grown from a warm-season and cool-season grass seed mix is evergreen. Besides such two types of lawn, the lawn grown by sowing flower seeds and grass seeds is

Figure 2.13   Grass dotted with flowers contributes to the park-like scenery along the Nanjing-Hangzhou Motorway.

dotted with colourful flowers, enriching the natural landscape. The lawn not only can grow on gentle slopes with high-quality soil and sufficient water content, but also can grow on very steep rocky slopes of the cliff (Figure 2.13). For those steep slopes, covering them with liana is also an effective biological slope protection measure. Most liana have adventitious roots, suckers or tendrils, with characteristics such as climbing, winding, hanging, etc., and it is capable of adapting to a variety of adverse environmental conditions, even extending up a 90° rocky slope. In addition, this type of plant grows fast which can cover bare slopes quickly, having a significant impact on greening, soil consolidation and slope protection (Figure 2.14).

Figure 2.14   The ivy planted on the steep slopes of the Nanjing-Hangzhou Motorway has covered the whole slope in less than a year.

Grass and ground cover plants for slope protection landscaping are commonly integrated with engineering protection measures, such as:

(1) For soil-filled slopes with a height of less than 3.5 m, grass or other ground cover plants can be directly planted on the slope after the gradient of the slope has been reduced (usually less than 1:1.5). Where the gradient is about 1:1.5, precast concrete hexagonal hollow blocks are first placed on the side slope of the subgrade, conical slope and sliding slope of the abutment as well as the subgrade slope behind the bridge, or masonry blocks and precast blocks are placed to form an arch layout, with precast sluice designed to divert water on the road surface into the ditch. Upon soil consolidation, the grass or ground cover plants are planted inside the hexagonal hollow blocks and arches. This is one of the methods most widely used in the filled subgrade slopes of the Nanjing-Hangzhou Motorway (Figures 2.15–2.18).

(2) For the steep section with a filling height of over 7 m, hierarchical construction of the retaining wall is first carried out, and a drainage ditch is built at the terrace to divide the whole slope into several small slopes at a height of around 3 m, the upper section with a gradient of 1:1.5 and the lower section with a gradient of 1:7.5. Then, grass is planted directly on the slope, or lush liana is planted to climb up or hang down the whole slope (Figure 2.19). Grass and ground cover plants are extensively used in side slope protection to create special landscapes which play a unique and important role in motorway landscaping. For example, the lower side slope of the Nanjing-Hangzhou Motorway is planted with a large amount of Bermuda grass or a combination of Bermuda grass and white clover.

(3) Since the Nanjing-Hangzhou Motorway goes through a large number of rocky hills, in order to reduce the influence of quarry waste on the

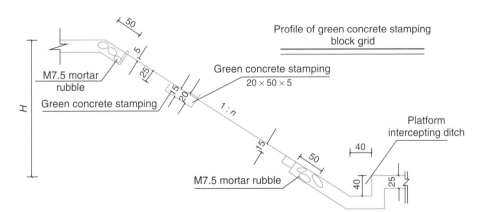

Figure 2.15   Profile of green concrete stamping block grid (unit: cm).

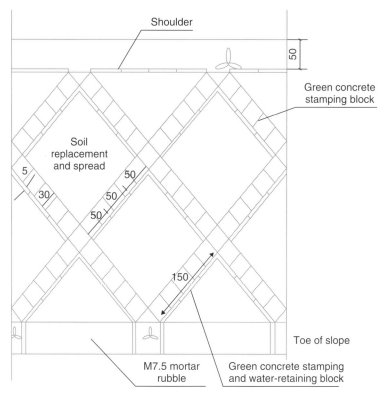

Figure 2.16   Slope protection design (unit: cm).

Figure 2.17   Planting grass inside precast concrete hexagonal hollow block.

Figure 2.18   Application of hexagonal hollow blocks on conical slope of overpass.

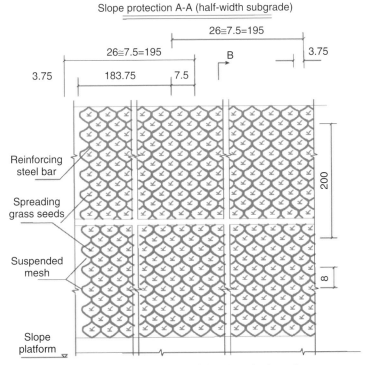

Figure 2.19   Design of suspended mesh of the slope (unit: cm).

environment and save project investment, excavated rock is used for filling the subgrade to form a rock-filled side slope. For the rock-filled side slope, a layer of about 1-m-thick clay is first placed on both sides, covering the edge with clay, and then planting grass and ground cover plants. With regard to the slope composed of large rock blocks from excavation, the blocks of rock are placed on both sides of the subgrade under dry conditions, using soil to fill in gaps in the rock before planting grass. As for the subgrade filled by gravel and sandy soil, grass can be directly planted on the slope (Figures 2.20–2.22).

Figure 2.20 Planting grass on a high subgrade slope near Donglushan Mountain.

Figure 2.21 Filling the subgrade with excavated quarry waste, covering with soil and planting grass on the slope to save costs.

Figure 2.22   Placing dry large rock blocks at the side of the subgrade.

The treatment of the upper and lower rocky side slopes is different. Appropriate engineering measures and greening technology are used according to the specific circumstances of the slope.

(a)   For steep but stabilized sections, after clearing away loose debris, soil replacement and spread against suspended mesh is carried out directly, or the 'grass stick' method is used first as a cushion, and then soil replacement and spread against suspended mesh shall be carried out, to cover the rocky slope with greenery.

(b)   For sections with gentle slopes and a certain depth of soil, soil replacement and spread shall be directly carried out after clearing away loose debris.

(c)   For the subgrade filled by gravel and sandy soil, spreading shall be directly carried out on the slope which will be covered by plants after seed germination and a period of growth.

As for the upper rocky slope, the following appropriate measures can be taken according to the specific circumstances of the slope.

(a)   For the steep but stabilized section, due to a lack of soil or poor soil quality, the basic conditions for plant growth cannot be met. Engineering measures are required for improvement, and soil replacement and spread against suspended mesh can be carried out for slope protection (Figures 2.23 and 2.24). Alternatively, waste tyres are fixed on the slope with bolts and tied together, then a mixture of plant seeds, fertilizer and

Figure 2.23   Slope dotted with colourful flowers along the Nanjing-Hangzhou Motorway.

Figure 2.24   View of golden *Hypericum monogynum* flowers on the side slope along the Nanjing-Hangzhou Motorway.

soil is filled into the tyres and gaps. This is consequently a good way to protect the environment after seed germination and growth.

(b)   For sections with relatively good soil quality and gentle slopes, soil replacement and spreading shall be directly carried out after clearing away loose debris. Alternatively, grass sticks made by wrapping straw on bamboo poles are fixed to the slope in a parallel manner by steel nails. Steel mesh is placed between grass sticks, and 4- to 5-cm-thick nutrient-rich soil is provided prior to spreading.

(c)   For the rocky soil section with relatively good soil quality and high soil content, holes or slots can be dug for planting. The holes or slots are filled with nutrient-rich soil to aid growth of the plants.

(d)   For the section with heavily weathered rock or partial instability, protection work is carried out first, with a reinforced concrete skeleton buried under the slope or fixed to the slope by bolts to improve the stabilization of the slope surface. Then suspended mesh soil replacement and spreading is carried out between the reinforced concrete skeleton.

Shrubs for protecting the slope landscape

As for the steep side slope with a gradient of less than 1:1 and filling height of over 5 m, a combination of engineering and biotechnical protection is adopted in order to protect the slope. Single or double layers of arch lining are first built onto the slope, and then one kind of shrub or various kinds of shrubs are planted inside the arch ring. The highly developed root system of these shrubs has a strong capacity in terms of soil consolidation and water retention, which can effectively prevent soil erosion and ensure beautiful and natural scenery. For example, on the lower side slope of the main road of the Ganma Service Area of the Fenshui–Guanyun Motorway, winter jasmine is planted on a 4–5 m high slope there. This kind of shrub is evergreen with beautiful yellow flowers in early spring (Figures 2.24–2.27). Another example is the interchange to the east of Lishui on the Nanjing-Hangzhou Motorway. Since the conical slope of the overpass and the side slope next to the road at the interchange are steep, a mixture of winter jasmine, greenstem forsythia and Chinese firethorn is planted in order to prevent soil erosion. The slope is covered with lush greenery after 3 months of growth. In early spring, little yellow flowers on the winter jasmine herald the arrival of spring, and the bright yellow flowers on the greenstem forsythia are equally dazzling. The small red berries of the Chinese firethorn in October represent the red colours of autumn, adding to the autumn harvest scenery (Figure 2.28).

Figure 2.25   Planting winter jasmine for slope protection.

Figure 2.26   Sound slope protection by winter jasmine.

Figure 2.27   Condensed planting of *Sabina komarovii* and *Berberis thunbergii* on the side slopetrimmed into patterns for decoration.

Figure 2.28   Mixed and condensed planting of greenstem forsythia, primrose jasmine and Chinese firethorn for slope stabilization.

On the side slope at the Luojiabian interchange area, a mixture of Spanish dagger and multiflora rose also ensures the protection of the slope (Figure 2.29). On the high subgrade steep slope of the Shangxing interchange overpass, four-level arches are formed by mortar and precast concrete blocks, with greenstem forsythia planted inside the arch for soil consolidation, which further improves the view of the slope together with Bermuda grass (Figure 2.30). Rhododendrons have been planted on the K42 slope, which look exceptionally charming when the flowers are in bloom (Figure 2.31). At the overpass side slope of the Taihu Service Area, a mixture of *Nerium oleander, Ligustrum ovalifolium* Vicaryi and China rose ensures protection of the slope, soil consolidation and landscaping (Figure 2.32). Besides, relevant data show that staghorn sumac have been planted on the side slopes of the

Figure 2.29   Mound lily at Luojiabian interchange area for protecting the landscape of the slope.

Figure 2.30   Protection of the high subgrade steep slope at the interchange overpass in Shangxing.

Figure 2.31   Side slope at the K41-K42 section of the Nanjing-Hangzhou Motorway is covered by red rhododendrons.

Figure 2.32   Effective mixed planting of various shrubs on the overpass side slope of the Taihu Service Area.

highway near Cangzhou in Hebei Province, where the whole slope has been quickly covered by them on account of the constant tillering of the trees, which spread at an annual rate of 1.2–2 m. In autumn the red leaves are a big attraction along the highway.

Mixed planting of grass and shrubs for protecting the landscape of the slope

After planting grass or ground cover plants on the slope, a certain number of shrubs are planted to form shrub and grass landscape, which is a relatively good way to achieve slope protection. Shrubs for slope protection cannot completely prevent soil erosion due to insufficient coverage in the initial stage, while a combination of shrubs and grass can avoid the above problem, highlighting a sense of hierarchy and strengthening the visual appeal. Planting flowering shrubs at one side or the edge or middle of the green grass is particularly appealing (Figures 2.33–2.35). For slope protection on the Nanjing-Hangzhou Motorway, various shrubs such as bamboo, greenstem forsythia and rhododendrons are planted on the grass. For the design of the Liyang section, Bermuda grass is planted on the grass of the slope, while Chinese rose, rhododendrons, greenstem forsythia, primrose jasmine, Christmas camellia, winter cassia and cottonrose hibiscus are planted near the road shoulder or toe of the slope. Large amounts of flowering shrubs at the ramped slope of the Taihu Service Area also ensure sound landscaping. In northern China a mixture of amorpha with buffalo grass works very well. According to reports, buffalo grass grows quickly and soon covers the ground, which

Figure 2.33   Grass and shrub structure.

Figure 2.34   Grass and shrub structure.

Figure 2.35   Grass and shrub structure.

will be able to prevent slope surface runoff and reduce transpiration intensity, thus successfully reducing erosion even in the initial stage. This improvement of the surrounding environment creates good conditions for the growth of the false indigo bush, and this plant has root nodules that can consolidate free nitrogen in the air, increase soil nitrogen and facilitate the growth of buffalo grass. Therefore, false indigo bush and buffalo grass complement each other's advantages to achieve good and lasting slope protection.

Combination of trees and shrubs for protecting the landscape of the slope

The combination of trees and shrubs for slope protection can achieve better ecological and landscaping effects. This kind of landscape design mainly applies in the following environmental conditions:

(1)   Strong visual demand for slope landscaping.
(2)   Long and steep slope prone to landslides, collapse or serious soil erosion.
(3)   Steep slopes on both sides of highway overpass.
(4)   Mountainous area prone to avalanches or landslides.
(5)   Gentle slope with a gradient of 1:5 or less.

For side slopes with different environmental conditions, slope protection measures involving the mixed planting of trees and shrubs should be taken for soil consolidation and slope stabilization as well as greening and landscaping, so as to improve and enhance the environmental surroundings of the road. However, the design should focus on different aspects according to specific sections. In cases (2), (3) and (4) above, soil consolidation and disaster prevention are highlighted. The main roots of trees and shrubs can go deep into the soil of the slope body, while the lateral roots and entangle roots become intertwined with the soil. In addition, the tree trunk is tall and straight, and the large crown of the shrub can prevent the slope from direct erosion by rain water, acting as an effective cushion. The mixed planting of trees and shrubs forms an ecologically and biologically complementary solution, with good soil consolidation and water retention effects as well as a strong three-dimensional sense. On the other hand, cases (1), (3) and (5) focus on creating a beautiful landscape. The planting of tall trees on both sides of the side slopes of the over-pass can visually shelter the overpass, improve the landscape on both sides of the overpass, and stabilize the side slope. Compared with the effect on slope protection, the planting of trees and shrubs on gentle slopes has a more significant impact on landscaping. The deep soil of the slope serves as a basis and guarantees the growth of woods. The wide space available facilitates the landscaping of the motorway. Therefore, the reduced gradient of the slope expands the area of greenery and creates the conditions for various natural landscapes, and therefore the view along the Nanjing-Hangzhou Motorway is extremely picturesque. The mixed planting of trees and shrubs on the gentle slope along the Nanjing-Hangzhou Motorway makes for a beautiful landscape, an effect hardly ever achieved on the other motorways in Jiangsu Province. Figure 2.36 shows the mixed planting of trees and shrubs on motorway side slopes in China and elsewhere, while Figure 2.37 shows the mixed planting of trees and shrubs on the side slope of the Nanjing-Hangzhou Motorway.

(a)

(c)

(b)

(d)

Figure 2.36   Landscape of the mixed planting of trees and shrubs on motorway side slopes in China and elsewhere: (a) trees on the slopes of a Japanese motorway; (b) trees and shrubs on the slopes of the Fenshui–Guanyun Motorway; (c) trees on the slopes of a Dutch motorway; and (d) trees and shrubs on the slopes of the Nanjing–Hangzhou Motorway.

(a)

(b)

(c)

(d)

(e)

(f)

(g)

(h)

Figure 2.37 (a–h) Landscape of mixed planting of trees and shrubs on the side slope of the Nanjing-Hangzhou Motorway.

Landscape of softening and green retaining walls

A retaining wall is often used to restrain soil in slopes. In engineering design, the retaining wall, perpendicular to or behind the road, is constructed from brick, stone, concrete, reinforced concrete and other materials. In the past, the design of retaining walls only focused on engineering safety, with less attention paid to landscaping effects. Therefore, the structure of retaining walls generally looks rigid and stiff, as if uncoordinated with the surrounding environment. When constructing highways with consideration to ecological, environmental and landscaping aspects, it is necessary to carry out softening treatment in the design of the retaining wall, provided that the protective function is still achieved. Green plants are used for greening and landscaping of the retaining wall, providing drivers with uniquely beautiful landscapes. According to Zhangyang data, the following forms can be adopted to design the landscape of the retaining wall:

(1) For the terraces with better soil quality and insignificant height difference, it would be best to avoid the construction of a retaining wall, and treat it as slope terrace to carry out greening. For the section with a significant height difference, terraced retaining walls can be constructed at the lower part, or multiple stone arches can be constructed to ensure soil slope stability and greening inside the arch (Figure 2.38).

(2) Retaining walls should be built by multi-terrace construction. Greening work should be carried out at the central terrace. This kind of multi-level sub-wall not only avoids a visually bulky and dull effect caused by large retaining walls but can also effectively improve the appearance of the wall using greenery (Figure 2.39).

(3) In the case of a significant height difference, the retaining wall facade can be divided into two parts, with the lower part wider to better stabilize

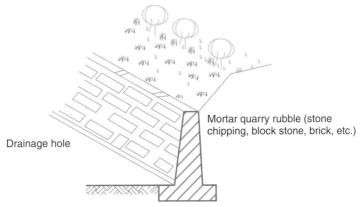

Mortar quarry rubble (stone chipping, block stone, brick, etc.)

Drainage hole

Figure 2.38   Terraced retaining wall at the foot in the case of big height difference.

the retaining wall. Planting slots or holes should be dug between the two parts for greening (Figure 2.40).

(4) The design of a vertical retaining wall into a sloped retaining wall can create a more open space and a brighter environment (Figure 2.41).

(5) Curved or broken line design is more attractive and enhances the appeal of the retaining wall whilst enriching the diversity of the landscape.

(6) Greening, embossed design and other treatment on bricks, stones, concrete blocks and other blocks or the retaining wall surface can improve the appearance of the retaining wall and improve and enrich its original landscaping effect. For example, under the retaining wall of mortar rubble on the Shanghai–Nanjing Motorway, Japanese creeper grow quickly and cover the whole wall, which is pleasantly green in spring and summer, with red leaves comparable with maple leaf in autumn, offering year-long beautiful scenery.

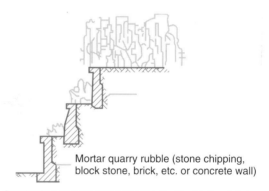

Mortar quarry rubble (stone chipping, block stone, brick, etc. or concrete wall)

Figure 2.39   Sectioning and layering design for the high retaining wall.

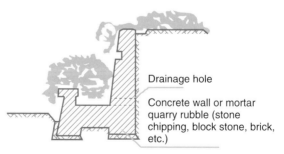

Drainage hole

Concrete wall or mortar quarry rubble (stone chipping, block stone, brick, etc.)

Figure 2.40   Retaining wall divided into two by planting groove in the case of big height difference.

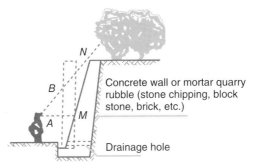

Figure 2.41   Vertical retaining wall designed to be inclined.

2.7   Selection of green plants for the slope

A slope environment is hostile for plant growth. Under such circumstances, it is essential to select plants that can adapt to this environment so as to green the slope. The following principles should be followed when selecting plants for the slope:

(1)   Native plants should be given priority, as they have good adaptability and strong resilience.

(2)   Grass should dominate in combination with shrubs and vines; trees should play a secondary role.

(3)   Use drought-resilient plants and plants resilient to infertile soil; choose plants that can grow and reproduce naturally.

(4)   Choose plants that reproduce via seeds and that are conducive to mechanical spray-seeding.

(5)   A certain amount of wild flowers and plants should be chosen and planted according to the landscape design requirements.

Plant species for slope protection in the territory of Jiangsu Province include:

(1)   Herbaceous plants: *Cynodorz ctylorz*, Bermuda grass, *Zoysia matrella, Zoysia japonica, Trifolium repens, Dichondra repens, Medicago sativa, Eragrostis curvula* and *Festuca arundinacea*.

(2)   Vines: *Parthenocissus tricuspidata, Hedera nepalenis, Parthenocissua tricuspidata, Vinca major, Euonymus fortune, Campsis grandiflora, Jasminum mesnyi, Jasminum mesnyi, Merremia boisiana, Trachlospermum jasminoides, Rosa multiflora* and *Jasminum nudirlorum*.

(3)   Shrubs: *Lespedeza formosa, Vitex quinata, Indigofera pseudotinctoria, Shibataea chibebsis* (*Sasa fortune, Indocalamus tessellatus*), *Cassia bicapsularis, Ligustrum quihoui, Pyracantha fortuneana, Amorpha fruticosa,*

Camellia sasanqua, Rhododendron simsii, Serissa japonica, Kerria japonica, Nerium oleander, Hibiscus mutabilis and *Camellia oleifera*.

(4)   Mega phanerophytes: *Ligustrum lucidum, Robinia pseudoacacia, Sophora japonica, Ailanthus altissima, Elaeocarpus decipiens, Cerasus serrulata, Prunus persica, Celtis sinensis, Pinus elliottii, Pinus taeda, Cedrus deodara, Zelkova schneideriana, Prunus cerasifera, Ginkgo biloba* L., *Acer palmatum, Melia azedarach, Firmiana simplex, Ulmus pumila* L. and *Populus euramevicana*.

2.8   Remaking of the rocky cut slope

The most difficult part of slope greening is the rocky slope where there is a shortage of soil and moisture for plants to grow. Carrying out construction on the steep slope is very complex and so is its maintenance. To solve these problems, many construction technologies and methods have been researched and developed. The rock slope in Donglushan Mountain is a good example; a brief introduction will now be given of its landscape design and construction technology in order to better understand the technical requirements of and methods for biological slope protection.

Donglushan Mountain is located in Lishui County on one side of the Nanjing-Hangzhou Motorway. In order to meet the route requirements, the sloped surface on one side of the mountain's peak had to be partially excavated (the volume of earth and rock amounted to approximately 400 000 m³). During the excavation of the mountain, all the original secondary vegetation was destroyed, including *Shibataea chibensis* (a species of bamboo) and *Pinus thunbergii* (Japanese black pine tree). The exposed rocky slope resulting from the excavation blast is 1460 m long and the exposed surface of the peak is 76.4 m high. In order to stabilize the weathered rock formations, prevent the falling of loose rocks, restore the damaged natural vegetation, reduce the amount of damage due to the construction of the motorway, and restore the natural environment to its original state, the following protection and greening methods have been designed for protecting the cut slope:

(1)   Engineering design of cut slope. Because the gradient and composition of soil and rock are different, many treatment methods are required for different slopes. The bottom of the slope features mildly weathered rocks; a slope ratio of 1:0.5 and top width of 2.5 m should be adopted. The middle of the slope features moderate to weakly weathered rocks; a slope ratio of 1:0.75 should be adopted; for the totally weathered rocks, the slope ratio should be 1:1. The first stage should be 5 m in height, the second 6 m in height and the third or above should be 8 m

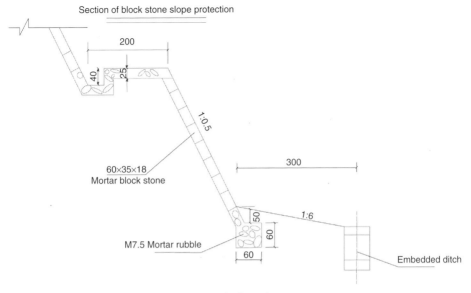

Section of block stone slope protection

200

40   25

60×35×18
Mortar block stone

1:0.5

300

M7.5 Mortar rubble

50

60

1:6

60

Embedded ditch

Figure 2.42   Engineering design of cut slope (unit: cm).

in height. The top should be covered with a layer of soil, with a slope ratio of 1:1.25 (Figure 2.42).

(2)   Protection and greening scheme for the cut slope. For the slope surface on the first and second stages, block stones should be used for overall protection. For the slope surface on the third stage or above, where the gradient is large, artificial materials should be used to conserve soil and vegetation by means of greening schemes, such as soil replacement and spreading against suspended mesh as well as bagging and wrapping grass seeds; for the top soil layer, grid skeletons should be added to consolidate the soil. On unstable blocks, anchor rods should be added as needed. In order to improve the greening effect, a planting trough is provided on top of every stage for planting trees, flowers and hanging plants; at the same time, the area between the bottom of the first stage slope surface and the subgrade edge is planted with trees and vines, enabling them to grow up the retaining wall (Figures 2.43 and 2.44).

(3)   The stage for heaping debris and planting vegetation. The stage on the retaining wall and ramp is a precious planting area, where there is a ditch for collecting the slope runoff and a retaining wall for preventing the loss of planting soil. For plants to survive and grow, it is advisable to keep the planting soil not less than 1 m in depth and width. In order to make three-dimensional landscapes, a combination of trees, shrubs

Small plants at the edge

Climbing plants

High
quality
surface
soil

Quarry
rubble wall

Figure 2.43   Greening of cut slope.

and vines can be selected. In order to ensure successful planting, tree species should be selected very carefully. Tree species should be small evergreen trees that are resilient to wind, infertile soil and dry conditions; shrubs can be evergreen or deciduous and should be resilient to wind and dry conditions; vines can also be evergreen or deciduous, and should be natural creepers as well as resilient to wind and dry conditions. In addition, the plants chosen should be able to adapt to minimal levels of management and grow naturally.

By adopting the above design and construction plans that take into account the actual conditions at Donglushan Mountain, the exposed hill surfaces have been covered with green plants; the entire peak is now green and attractive and the previously adverse environment has been totally improved. Meanwhile, the abandoned quarry has been filled with earth and changed into an area of grass, which is covered with paths and landscape rocks. A viewing pavilion has been constructed, providing travellers with a place for sightseeing and relaxing. Hence, this previously unattractive rocky area has been transformed into a picturesque scenic spot (Figures 2.45–2.51).

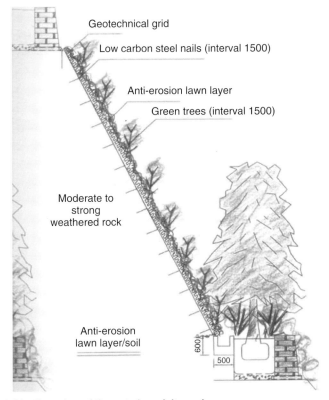

Figure 2.44   Greening of the cut slope (uit: mm).

Figure 2.45   Construction in progress.

Figure 2.46   Recessed surface is filled with straw bags.

Figure 2.47   Re-greening technology by spreading against suspended mesh on the Donglushan Mountain slope after multi-stage treatment.

Figure 2.48   Side view of each level of the stage for heaping debris and the draining ditch.

Figure 2.49   Re-greening of abandoned quarry.

Figure 2.50   Paths and viewing pavilion are built on the rocky ground after re-greening.

Figure 2.51   The construction of a pavilion and the arrangement of landscape rocks have created an attractive sightseeing spot.

2.9 Slope greening and construction technologies

Soil replacement and spreading

(1) Engineering principle. By applying the theories and methods of ecological greening engineering, this technology is used to restore the structure and natural functions of vegetation by artificial means. First, the soil structure is improved before uniformly spraying a mixture onto the slope surface using a hydroseeder. The mixture, generally known as 'soil replacement', contains many sorts of plant seeds, local good quality soil, fertilizers, adhesive and soil stabilizing agent. As the seeds sprout and grow, the exposed slope surfaces become covered with plants and the slope becomes green. When planting vegetation, a combination of pioneer, intermediate and target plants should be considered. Native wild plant species should be selected as far as possible; in addition, a certain proportion of fast-growing shrubs and small trees should be planted to ensure the long-term stability of the plant community on the slope. By using this kind of technology, plants are able to grow on the rocky slope, reinforce the slope and improve the motorway landscape. Moreover, the plants do not need intensive management, thereby saving on manpower and material resources.

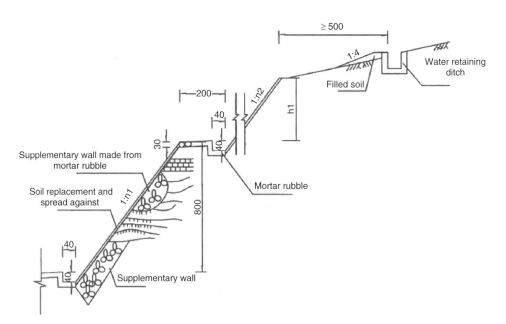

Figure 2.52 Construction diagram of soil replacement and spread against suspended mesh (unit: cm).

(2)   Construction technologies. There are two types of soil replacement and spread: ordinary and suspended mesh. The former involves directly spraying seeds onto the soil slope. This is a relatively simple process. The steps for soil replacement and spread against suspended mesh are as follows (Figure 2.52):

(a)   Clear and renovate the slope surface. Clear away debris and large rocks from the slope surface. Round off sharp edges and build a supplementary wall in areas of recess to ensure the slope surface is even (Figure 2.53).

(b)   Suspending mesh. Cover the slope with #12 or #14 galvanized lozenge wire mesh (diameter of wire 2 mm, mesh size 40 mm × 40 mm). Use reinforcing bars anchor bolts (the main anchor rod is 16 mm in diameter and the supplementary anchor bolt has a diameter of 12 mm). Bore holes perpendicular to the slope surface, place main anchor bolts at an interval of 200 cm, then arrange the supplementary anchor bolts exactly half way between the main anchor bolts (100 cm). Drill the anchor bolts 20 cm into the ground, leaving 20–40 cm out of the ground, or secure the anchor bolts with mortar and then suspend the mesh on the anchor rods before securing firmly (Figure 2.54).

(c)   Requirements for spread material. Ensure the spreading seeds are of good quality. Before spreading, determine the sprouting rate of the seeds; accelerate germination of the shrub seeds, but the shoots should not be too long so as to avoid damage to them during stirring before spreading. Select compound fertilizer containing nitrogen, phosphorus and potassium in a ratio of 15:15:15 or 10:10:5. Choose plant fibre that can absorb and retain water as the covering material. The dose of binding agent should be at least $1.0 \, kg/m^3$, while that of the water-retaining agent should be at least $2 \, kg/m^3$ (Figure 2.55).

Figure 2.53   Photograph of clearing up and levelling slope surface.

Figure 2.54   Photograph of suspending mesh.

Figure 2.55   Photograph of spreading seeds.

(d) Ratio of shrub and grass seeds. Based on the features of the climate and rocky slopes along the Nanjing-Hangzhou Motorway, the ratio of shrub and grass seeds should conform to the following requirements: *Lolium perenne* 0.17 g/m^2, *Festuca arundinacea* 0.29 g/m^2, Bermuda grass 0.58 g/m^2, *Trifolium repens* 1.67 g/m^2, *Medicago sativa* 0.58 g/m^2, *Magnolia multiflora* 3.89 g/m^2, *Lespedeza formosa* 5.93 g/m^2, *Pyracantha fortuneana* 1.27 g/m^2, *Mirabilis jalapa* 0.50 g/m^2, *Tagetes patula* L. 0.50 g/m^2.

(e) Mix soil with other matter. Select various organic matters, such as crushed mushroom, husk and sawdust, fertilizer, binding agent and water-retaining agent before adding to the soil and mixing well.

Figure 2.56   The slope is covered with film.

In order to achieve uniform spreading and accelerate seed germination, mix the above materials in the hydroseeder for 20 min and then spread.

(f)   Cover the slope with film. The slope should be covered with film from top to bottom with an overlap of 10–15 cm, securing with wooden or bamboo sticks. The film should be at least 120–180 g/m^2 (Figure 2.56).

(g)   Sprinkle water thoroughly. Water should be sprinkled thoroughly; however, the water should not cause water loss or runoff. It is important to prevent the washing away of substrate matter.

(h)   Tending and management. The surface should be covered with nonwoven fabric 2–3 days after spreading; 30–40 days later, uncover the fabric to harden-off the seedlings. During the tending period, water the surface appropriately according to the soil moisture content and seedling and weather conditions. Too much water may cause the soil to become compacted and suffocate the seedlings. As the replaced soil contains organic matter and efficient compound fertilizer, there is normally no need to apply additional fertilizer (Figures 2.57–2.60). Figure 2.61 shows the scene after successful revegetation.

Reinforcing the soil replacement and spread with continuous fibre

In recent years, Japan has succeeded in planting vegetation on slopes that have few fissures. These slopes are problematic because it is difficult for roots to become established in the ground. Japan succeeded by reinforcing the soil replacement and spread with continuous fibre. This method involves combining soil replacement, plant root systems, glass fibre and metal mesh to form a

Figure 2.57   The seedlings are hardened off.

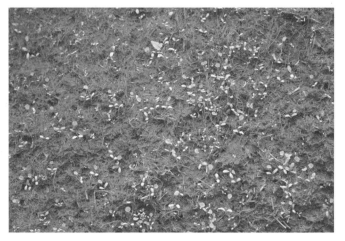

Figure 2.58   Seedlings after removal of the film.

Figure 2.59   Turfing.

Figure 2.60   Watering.

Figure 2.61   Successful revegetation.

three-dimensional mesh, which is then used to cover the surface of the slope and secured onto the slope by anchor bolts. The detailed technical process involved is as follows:

(1)   Clear up and renovate the slope surface.
(2)   Drill and insert anchor bolts into the surface. The anchor bolt is made of twisted steel, with one end welded with a small steel plate of 5 cm×10 cm. The anchor bolt is then drilled and inserted about 20 cm deep into the slope (Figure 2.62).
(3)   Lay down plastic drainage boards. Plastic drainage boards are generally made of waste plastic. They are placed at a gradient of approximately

Figure 2.62 Drilling and inserting anchor bolts.

25° to the vertical and secured with steel nails. Their main function is to drain surface water from the slope surface and fissure water in the slope (Figure 2.63).

(4) Spray sandy soil and continuous glass fibre towards the slope surface (Figure 2.64) with a thickness of approximately 20 cm. As glass fibre results from the flow of water, it is distributed evenly and naturally. Operation of this new innovation is very straightforward.

(5) Continue with soil replacement and spread against suspended mesh.

Figure 2.63 Laying draining pipes.

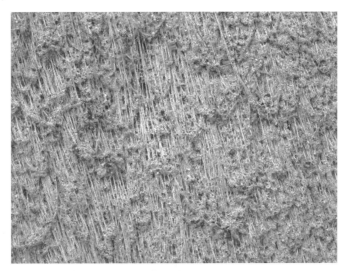

Figure 2.64   Spraying mortar and glass fibre.

Technology of soil replacement combined with straw sticks

Planting grass by applying soil replacement and spread combined with straw sticks is also an effective re-greening method. The detailed operating steps involved are as follows:

(1)   Renovate the slope surface. Remove debris and smooth down the surface. Fill recess areas with supplementary wall.

(2)   Underlay construction. Treat the straw with agricultural pesticide; bind the straw into bundles with a length of 4 – 5 m and soak the bundles in mud; secure them with steel nails (18 mm in diameter) and ensure a 20 cm gap between bundles and a 1.5 – 2 m gap between steel nails. The straw sticks should then be used to cover the surface of the slope. The bundles should be parallel to the motorway surface. After securing all of the bundles, cover them with soil 7 – 8 cm in thickness before smoothing over.

(3)   Suspend the mesh. Use #12 or #14 galvanized lozenge wire mesh (diameter of wire 2 mm) and secure it with steel elements.

(4)   Spread seeds. After watering, spread seeds; after spreading, cover the slope surface with nonwoven fabrics; sprinkle water thoroughly and cover the surface with mulch films.

The germination ratio of plant seeds can reach in excess of 85%. One month later, the rate of vegetation coverage is 80%; 3 months later this figure is 90%. Figure 2.65 shows the sequence of replanting.

(a)

(c)

(b)

(d)

Figure 2.65   (a–d) Sequence of replanting of grass via straw sticks.

Three-dimensional vegetation net technology

(1) Engineering principle. The three-dimensional vegetation soil consolidation net is a three-dimensional net structure composed of multiple layers of convex and concave nets and high-strength flat iron wire nets. The geogrid is an example of such a net. Its surface layer looks rough and uneven and its material is loose and flexible. More than 90% of its spaces are used to accommodate soil or a mixture of soil and grass seeds. Research has proven that before the appearance of turf, the soil consolidation rate of the three-dimensional vegetation net is 97.5% when the gradient is 45°; when the gradient is 60°, its soil consolidation rate is 84%; and when it is 90°, the rate is still 60%. It is therefore evident that the effect of the three-dimensional vegetation soil consolidation net is much better than other equivalents. In addition, because its surface is rough and uneven, it can effectively prevent grass seeds and their seedlings being washed away by rain water, thereby greatly enhancing the rate of vegetation coverage. After the grass has grown into turf, its soil consolidation rate can reach 100%. This is because the plant root system can penetrate the net, growing to over half a metre into the ground. This ensures a secure formation together with the net and soil. In addition, the plant root system can increase the water permeability of the soil. As soon as there is rain, water can rapidly permeate the soil. Vegetation coverage

can also protect the surface soil against the direct impact of rain water, and reduce the flow of rain water. Meanwhile, the three-dimensional net and the plant root system can function as reinforcements for the superficial layer. This kind of technology can therefore effectively protect the slope. The three-dimensional net is made of polyethylene, so it is chemically stable and nontoxic. Underground, its service life can reach in excess of 50 years, and even if it is exposed to direct sunlight, its service life is around 10–20 years. In addition, degradable plastic nets can also be used. After several years the plastic degrades and disappears.

(2) The geogrid three-dimensional net construction process is as follows:

(a) In the same way as soil replacement and spread, suspend the galvanized mesh on the slope surface.

(b) Lay the geogrid three-dimensional net (Figure 2.66) on the slope from top to bottom and secure with iron nails at 1 m intervals to ensure contact between the net and the ground. For the exposed or irregular areas, use more iron nails to secure the net, so as to prevent too many gaps underneath the net. Ensure an overlap between nets of 150 mm, making the net further up the slope to cover the net further down.

(c) Place soft soil in the net and compact and fill the voids.

(d) Evenly spray the prepared hydraulic spread materials (plant seeds and spread carrier matrix) on the geogrid three-dimensional net.

(3) Select seeds and seeding time. Select seeds appropriately according to the natural conditions, such as soil property and climate. In general, the seeds should meet the following conditions:

(a) Not requiring high-quality soil, saline and alkaline tolerant, resistant to infertile soil.

(b) Adaptable to different climatic conditions, such as extreme temperatures and drought.

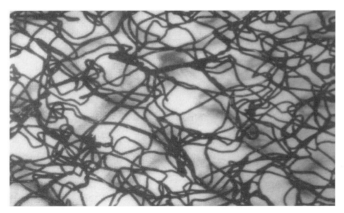

Figure 2.66   Three-dimensional vegetative net — geotechnical grid sample.

(c) Developed root system, ample foliage, fast growing, and long green period.

(d) Convenient to purchase and low cost.

In practice, grass seeds are normally combined. One type involves the 'pioneer' seeds of the grass family, such as *Festuca arundinacea* and *Lolium perenne*. This type of grass grows fast and can cover the ground surface quickly. Within 40 days it can cover over 80% of the slope surface. Another type of grass seed is *Coronilla varia*, which has many branches and grows across the surface. It grows to over 1 m long and its roots reach over 0.5 m deep. It is also resilient to weeds and produces many seeds. After its seeds fall onto the ground they can grow naturally. It is perennial, with a creeping stem, subterraneous stem, and a strong division between its root and stem. For this reason it can secure itself well into the ground. Its green period is approximately 300 days and it has excellent tolerance to cold temperatures. However, it grows slowly, taking 2 years to 'outgrow' the pioneer grass, which then acts as a fertilizer.

If shrubs are mixed with grass, the effect is even better. The grass grows rapidly, quickly covering the surface of the soil. Even during early growth it can already intercept surface runoff on the slope surface and therefore mitigate erosion. Shrubs such as *Amorpha fruticosa* grow slowly at first and cannot cover the surface very well, but they can protect the slope for a long time. Grass and shrubs therefore complement each other well and provide an effective and long-lasting protective result.

Planting pit technology

For the slope sections that have good-quality soil, thick layers of soil or stony soil, a planting pit or trough can be dug to plant shrubs, bamboo and ground cover plants. This enables the construction of a green multi-layered slope protection landscape that easily blends into the surrounding environment. Along the Nanjing-Hangzhou Motorway route there is a particularly large amount of bamboo, the planting of which enables the motorway to fit better into the surroundings (Figures 2.67 and 2.68).

(1) The materials for planting should be conducive to the growth of uniform, healthy and strong plants unaffected by pests. Native bamboo should dominate, the height of which should be the same as far as possible. Each cluster should have about 10 bamboo plants including bamboo rhizomes and earth ball.

(2) Remove debris and loose rock from the slope. Renovate the slope, fill recess areas with earth before compacting or fill with grass, resulting in a flat surface.

(a)

(b)

Figure 2.67   (a, b) Undershrub planted in the dug pit.

(3)   The planting pit or trough should be dug based on the root system, earth ball and soil conditions. The specifications are shown in Table 2.4.

(4)   Backfill soil. Before planting, nutrient soil should be backfilled into the pit or base fertilizer should be applied.

(5)   Requirements for planting. The distance between plants as well as their height should be determined according to the design requirements. The roots should be untangled and the backfilled earth compacted in layers. For shrubs and bamboo, the required 1-month survival rate is 85%, whereas the 1-year survival rate should be 90%.

Figure 2.68   Bamboo planted in the dug trough.

Table 2.4   Dimensions of shrub planting pit and bamboo planting pit.

Dimensions of shrub planting pit		
Crown diameter (cm)	Depth of planting pit (cm)	Diameter of planting pit (cm)
100	60~70	70~90
80	50~60	50~70
60	40~50	40~50
50	40~50	40~50
40	40~50	40~50
Dimensions of bamboo planting pit		
Depth of planting pit (cm)		Diameter of planting pit (cm)
30~40 deeper than packing or earth ball		50~60 wider than packing or earth ball

Vegetation bag technology

Vegetation bags may be divided into two types according to their application:

(1) Seed vegetation bags (Figure 2.69). Evenly sow a certain proportion of base matrix containing seeds (grass seeds, flower seeds and shrub seeds) and organic fertilizer in the two-layered nonwoven fabric. Connect the nylon protective screening with the nonwoven fabric containing seeds by means such as seaming, needling and gluing. Then spread the mesh bag flatly on the soil slope, securing the upper and lower edges in the soil.

Figure 2.69   Seed vegetation bag.

Cover it with a layer of cultivated soil 1–2 cm in thickness and wait for emergence of seedlings after watering.

This method is suitable for hard soil slopes, sandy soil slopes and heavily weathered soft rocky slopes. The slope length should be between 3 and 5 m and the gradient between 1:1 and 1:1.25. It is appropriate for both upper and lower slopes.

The advantages are: seeds are sandwiched between soluble nonwoven fabric layers to prevent them from being washed away by rain water. There is slow-release fertilizer in the bag, providing the plants with necessary nutrients for their growth. Furthermore, it is light, easy to implement and handle and has a good greening effect (Figure 2.70).

Figure 2.70   Effect on construction site of seed vegetation bag.

(2)   Individual vegetation bag. Put seeds (selected according to the design requirements, including flower seeds, shrub seeds, grass and tuber), organic fertilizer and cultivated soil into a green mesh bag with an inner layer of nonwoven fabric and an outer layer of PVC. Then seal its opening to form an individual object (Figure 2.71). This vegetation bag may be used as a component of a supplementary wall or for greening steep rock surfaces and rocky slopes with a gradient approaching 90° (Figure 2.72). Provided it is watered thoroughly, the seeds within the bag will sprout and grow, thereby achieving effective results.

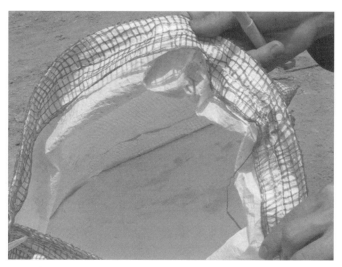

Figure 2.71   Single vegetation bag.

Figure 2.72   Vegetation bags piled up at various angles to fill in the rock surface.

Tyre slope protection

Waste outer automobile tyres can be placed onto the rocky slope and tied together with galvanized iron wire before being secured with anchor bolts. Then they can be filled with cultivated soil or replaced soil consisting of plant seeds, fertilizer and soil. Thus, turf, shrub-grass community or small trees may be planted as per the design requirements. This method recycles waste resources and is effective in both reinforcing the slope and revegetation. It is used widely in many regions in China. It is also demonstrated to good effect on the slopes along the Nanjing-Hangzhou Motorway (Phase 1) (Figures 2.73 and 2.74).

Figure 2.73   Laying tyres on the slope surface.

Figure 2.74   Effective slope reinforcement and revegetation by using tyres along the Nanjing-Hangzhou Motorway.

3 Interchange Environmental Landscape Design

3.1 Concept

Motorway interchanges are generally two-layer structures and serve as an important part of traffic hubs as well as entry and exit points for cities and regions. Interchange areas are separated by ramps into several areas of different shapes and sizes. The total area is very large, ranging from tens of thousands of square metres to hundreds of thousands of square metres. On the Wuxi–Jiangyin Motorway section, the Wuxi Interchange Xuzhou Hub and East Xuzhou Interchange each boast a total area of over 200 000 square metres (Figures 3.1 and 3.2). Interchanges are classified into single-trumpet and double-trumpet interchanges according to different traffic needs. As double-trumpet interchanges occupy a large area, toll stations and management centres are often built. In the interchange area, there are structures and traffic facilities such as flyovers, ramps and high road lamps. It occupies a large area, there are many topographical changes and there is far-reaching landscape, providing drivers with an attractive view. Expansive areas such as these are therefore an important part of motorway landscape design. On motorways in countries with well-developed traffic infrastructure, there are often several to dozens of lanes, meaning that the interchange covers a larger area and features more complex and multi levels (Figure 3.3).

3.2 Landscape design objectives

The main objectives of landscape design for the interchange area are given in the following.

The Environment and Landscape in Motorway Design, First Edition.
Qian Guochao, Tang Shuyu, Zhao Min and Jing Chun.
© 2014 China Communications Press. Published 2014 by John Wiley & Sons, Ltd.

Figure 3.1   Wuxi Interchange on the Wuxi–Jiangyin Motorway.

Figure 3.2   Xuzhou Hub Project at the intersection of the Xuzhou–Suyu Motorway and the Xuzhou–Lianyungang Motorway.

Figure 3.3   Interchanges of countries with advanced traffic infrastructure have more levels.

Guide drivers' sight

On the outside of the ramp, a green belt with tall trees can guide drivers during the journey and help them predict how much the ramp bends. Trees can also relieve drivers' mental tension and sense of danger when driving on high ramps. Even if a vehicle comes off the ramp, trees can reduce damage caused in an accident. In addition, at the triangular diversion end point, low shrubs and climbing vine plants not only have a physical buffering effect but they can also prompt drivers from a distance and guide them along the route (Figures 3.4 and 3.5).

Figure 3.4   Example of guiding drivers' sight.

Figure 3.5   Example of guiding drivers' sight.

Protect the slope and consolidate the soil

The slopes at both sides of the ramp as well as the conical slopes at both ends of the flyover are often steep and very prone to water and soil loss, resulting in damage to the pavement. In the past, this problem was solved by using mortar rubble. However, the rubble does not blend well into the surrounding environment. An improved method is to combine engineering measures with biological measures or use only biological measures. When grading the slopes of the Nanjing-Hangzhou Motorway, the biological measures applied have ensured the protection of the slopes and the consolidation of the soil and improvement of the landscape. For example, flowering shrubs are dotted amidst the grass planted on the slope, not only improving soil consolidation but also improving the appearance of the environment. The originally stone-paved and unattractive landscape has been transformed into a green space full of blooming flowers (Figure 3.6).

Plant vegetation and improve the motorway environment

Tens to hundreds of thousands of square metres of land within an interchange area is a wonderful site for motorway landscape construction. By taking advantage of the local natural and cultural environment and combining this with motorway traffic needs, many different themed landscapes can be created along the route using different forms of art. These landscapes are a highlight of the motorway route, which enhances the quality of the surrounding landscape, providing picturesque scenery for drivers along the route. Figure 3.7 shows the Zhongshanmen Interchange on the Nanjing–Shanghai Motorway. The landscape design of this interchange accentuates the Nanjing

Figure 3.6   Plants both consolidate the soil and improve the environment.

(c)

Figure 3.7   **The Zhongshanmen Interchange: (a) bird's-eye view of the interchange; (b) statue of a magical beast; and (c)** *Prunus mume* **in bloom.**

natural scenery, integrating the surrounding mountains, rivers, city and forest. It also reflects the ancient city's large number of scenic spots and historical sites. The combination of the statue of a magical beast, *Cedrus deodara*, *Plotinus hispanica* and *Prunus mume* found in the interchange area is a testament to the ancient city's nature and history.

3.3   Creative landscape design

Landscape design shall be based on comprehensive analysis on local physical geographical characteristics, economic and social development, customs and habits, historical and cultural heritage, terrain features within the area, buildings and functions of transportation, etc., for creating green landscapes with unique ecological and cultural features.

Partial greening design of motorway interchanges around China presents strong local cultures. For example, Xingping Interchange of the Xi'an-Baoji Motorway is designed using the patterns 'Imperial Concubine Yang appreciating jade' and 'Huo Qubing (a General in Western Han Dynasty) observing the inverted Big Dipper', referring to the local historical legends about Concubine Yang and Huo Qubing, enriching the historical interest for the entire landscape and simultaneously promoting the abundant local tourist resources. Yunjiakou Interchange in the eastern suburb of Xining is another example. The interchange is designed with a crescent pattern to reflect the Islamic culture catered for by the well-known Grand Mosque nearby. There are many cases of cultural features being applied to interchange designs in different regions in China, including three interchanges in the Gantang–Wengcheng section of the Beijing–Zhuhai Motorway designed to reflect local different plants: *Erythrina* blossoming in winter displayed at the Qujiang Interchange; *Michelia*, a local well-known flowerat the Shaxi Interchange; and green Cedar at the Wengcheng Interchange. These all serve as representations of local geographical features, customs and habits. In particular, on the Guangdong Yangmao Motorway put into operation in 2004, there are several interchanges designed to embrace local conditions in a diversified manner, highlighting waterscape or forests. For example, Chengcun Interchange is dominated by groups of big trees, Zhikui Interchange presents an orchard style, Xinxu Interchange shows a grassland effect combined with tall and straight *Ceiba*, Mata Interchange looks like a golf course with a mixture of water and grass and Guanzhu Interchange expresses a coastal landscape consisting of beach, reed and Hainan coconut.

In Jiangsu Province, Qintong Interchange on the Jinjiang–Yancheng Motorway can be used as a good example. The interchange, located in waterland conditions and running through fish ponds, is designed to highlight the waterland features and local traditional water activities. According to the design, a pond is excavated in the low-lying land of the interchange and then the excavated earth is borrowed for filling the wet land beside the pond to shape a slope; lotus flowers planted in the pond and willows along the pond create an illusionary and looming vision against weeping willows; on the slope, undershrubs are made into mould patterns to show the grand occasion of 'competition among thousands of boats sailing on the great Qin Lake'.

3.4   Main types of landscapes

The interchange landscape designs are diversified. To sum up, there are the following several categories.

Rural category

Some interchanges are located in the countryside, far away from urban areas. Plants are distributed naturally, emphasizing the natural landscape of the region and highlighting the layering and three-dimensional effect of greening and ensuring that the interchange blends well into the countryside (Figures 3.8 and 3.9). Natural ecological greening is currently the main form of the interchange greening. The rural category can be divided into the following subtypes:

(1) Sparse woodland and grassland. Within interchange areas there are natural ups and downs in the landscape, or there are artificial pools or earth mounds, where a landscape of sparse woodland and grassland can be designed on the slope or earth mound. Trees and grass can be used to build natural woodland with flower-dotted grass. This ensures the creation of a natural, open and peaceful landscape (Figures 3.10 and 3.11).

Figure 3.8   Countryside interchange landscape.

Figure 3.9   Countryside interchange landscape.

Figure 3.10   Sparse woodland and grassland type.

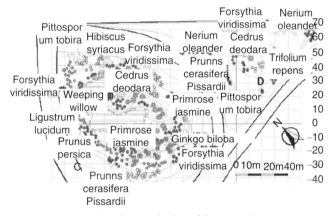

Figure 3.11   Interchange landscape design of forest and grass.

(2)   Rural scenery. There are toll stations, management centres and staff living quarters in some interchange areas. This type of interchange often covers a large area. In areas of residential development, natural garden landscapes that fit in with the modern buildings can be built. This creates an environment that is suitable for both living and working in. Around its periphery, vegetation should be planted that simulates the local natural vegetation and a large area of woodland should be built to reduce traffic noise, dust and exhaust pollution (Figure 3.12).

(3)   *Jiangnan* (area south of the Yangzte River) waterside scenery. Within the interchange area, there are large ponds, lakes or rivers, so a landscape characteristic of Jiangnan scenery can be designed according to the local conditions. First, the waterways should be adjusted such that they link up with the drainage system to facilitate the drainage of rainwater during

Figure 3.12   Rural scenery.

Figure 3.13   (a–f) Interchange landscapes of rural and waterside scenery.

the rainy season. Floating plants and emergent aquatic plants are arranged on some of water surfaces, and hygrophilous trees and shrubs are planted on the bank, forming a varied layout of plants that is both high and low and sparse and dense. Nearby the subgrade under the slope some small shrubs are planted (Figure 3.13).

(4) Aquatic and wetland. In regions where there are many rivers, lakes and wetlands, some motorways pass across lakes. There are therefore good reasons to build wetland-based landscapes in these environments. According to the height of the water, hygrophytes such as *Phragmites communis, Zizania caduciflora, Typha angustifolia, Acorus calamus* Linn., *Juncus effuses* and *Scirpus tabernaemontani* can be planted over large areas. In addition, moisture resistant trees and shrubs, such as *Taxodium ascendens, Ascendens mucronatum, Taxodium distichum, Pterocarya stenoptera, Ilex chinensis, Salix babylonica* and *Populus tomentosa* can be planted on the bank, ensuring a good landscaping effect (Figure 3.14).

(5) Woods. A large area of both evergreen and deciduous tree and shrub species may be planted in the Q-shaped area of the interchange area to form a wooded landscape. It is not advisable to arrange large trees within the triangle area due to traffic safety, therefore shrubs and vine plants should be planted here. Large orchards and nursery gardens may also be built (Figure 3.15). In some cases, interchanges are located beside a forest. In this case, it is beneficial to introduce the forest to the interchange, seamlessly integrating the motorway into the natural environment (Figure 3.16).

Suburb category

Some interchanges are located on the outskirts of a city or within a city itself. They often cover a large area. A combination of measures to green

Figure 3.14   Wetland landscape within the Shangxing Interchange on the Nanjing-Hangzhou Motorway.

Figure 3.15   Large orchard and nursery garden within the interchange zone.

Figure 3.16   Large area of poplar trees within the North Xuzhou Interchange forms a wood.

and improve the appearance of the interchange should therefore be taken to ensure that it blends well into the surrounding urban buildings and green landscape. The interchange should also reflect the local cultural and social environment. For example, the North Wuxi Interchange is an important entrance to the Shanghai–Nanjing Motorway and also the only way which must be passed from Jiangyin to Wuxi. Covering an area of 30 hm^2, its green landscape is designed according to factors such as the local environment, traffic conditions and historical background. Given that vehicles travel fast on motorways and therefore pass by landscapes very quickly, interchange landscape design focuses on a simple yet noticeable and lasting effect. For this reason, existing ponds are reconditioned and enlarged to highlight the water-rich landscape. On the pond bank trees such as *Cinnamomum camphora* and *Metasequoia glyptostroboides* are planted as well as wetland plants such as reeds and weeping willows. A flower and grass landscape is employed on the edge of the woods. A vivid Jiangnan waterside scene is therefore created for visitors. The whole green landscape is both natural and inviting, attracting the interest of many visitors.

Suburb interchanges may be divided into the following subtypes:

(1)   Pattern type. Within interchange areas near urban areas, patterns may be made to blend into the urban green landscape (Figure 3.17). The land can also be developed into a green area for advertisements in order to promote enterprises, goods and tourist attractions. However, the pattern should be simple and conducive to care and management.

(2)   Cultural landscapes. In a city's surroundings, there are often cultural landscapes, such as scenic spots, historical sites and temples. Interchanges adjacent to these scenic spots may be made with abstract patterns to reflect the local history and culture, and therefore make a contribution to local economic and social development (Figure 3.18).

Figure 3.17   Vegetation pattern.

Figure 3.18   Culture-oriented landscape.

Figure 3.19   Composite landscape.

Atypical category

This refers to neither rural nor suburb types, both of which may be combined freely. Vegetation patterns can be added to the natural surroundings, and naturally sparse trees and shrubs can be planted within the patterns, forming a composite landscape (Figure 3.19).

In order to create the landscapes mentioned above, the natural topography must be reconditioned. Ponds should be linked to the side ditch and draining system. Pond banks should not generally be constructed with stone, nor built with dry rivers, but rocks can be used for decoration. The combination of plants on the bank should be chosen naturally and appropriately. Weeds and wild flowers that do not require intensive management can be used to make the grassland more attractive, and artificial grassland may be used accordingly. The creation of scenic spots should fit in with the local environment and they should be convenient to manage in the future.

3.5   Design features of interchanges on the Nanjing-Hangzhou motorway

The Nanjing-Hangzhou Motorway (Phases I and II) between Luojiabian in Lishui and Tizishan Tunnel in Yixing is more than 100 km long. There are 11 interchanges, each of which is designed with a landscape according to the 'pearl necklace' concept. According to the concept, each interchange is seen as a 'pearl' inlaid on the long 'chain necklace'. Each 'pearl' is carved according to different topographical conditions and the local natural and cultural landscape in order to determine a theme. Unity of the style of the entire route is first ensured before accentuating the features of each 'pearl' and making each one 'shine', providing the motorway route with scenic landscapes. At the same time, attention is paid to people-oriented design and natural ecology, avoiding damage to the natural environment. The mental and visual impact on drivers and passengers is also taken into consideration so as to create a comfortable and attractive environment surrounding the motorway. On the Nanjing-Hangzhou Motorway, artificial landscapes have been seamlessly integrated into the surrounding countryside. Drivers and passengers are therefore able to experience beautiful landscapes. Each interchange is unique and leaves a lasting impression.

Lowering of the gradient of all slopes within the interchange

The slope ratio is generally lowered to 1:5 or less, and in some places the slope is almost parallel to the pavement. By reducing the gradient of the slope and removing conventional mortar rubble, the green area within the interchange area can be enlarged without perception, making the interchange both more spacious and flatter. Drivers and passengers can have a safer view of the landscape from a gentle slope than from a steep one. At excavated sections over 200 m in length and filled sections with a subgrade height of less than 1 m, the crash barriers are removed and replaced with contour lines. Removing the crash barriers makes the motorway appear more natural and spacious (Figure 3.20).

Distinct themed design of each interchange

These themes are determined by analysing and comparing elements such as the local natural and geographical features, folk customs, history and culture as well as traffic and transportation needs. Each interchange has its own distinct regional and ecological features.

As an example, the Luojiabian Interchange is located at the scenic spot of the Tiansheng Bridge and the romantic Yanzhi River and features famous natural scenery (Figure 3.21). The designer arranged plots of single, tall trees within the interchange area to form a smooth winding pattern. This was selected as the main way to design the surrounding forest and can be seen in many areas within the interchange. Various kinds of plants are grouped together in a winding pattern, just like a flowing river and a wave moving up and down in the wind. The themed design is remarkably lifelike. It is worth noting that the design of this interchange also takes the peripheral area into account, ensuring that it blends into the main scenic spot. For example, a green belt composed of the same plants used within the interchange area has been placed outside the approach. These plant landscapes not only integrate

(a)

(b)

Figure 3.20   (a, b) A large number of trees and shrubs are planted on the graded slopes of the Guizhuang Interchange.

with those within the interchange area but also fit into the surrounding area. This man-made interchange landscape has therefore blended into nature and become one of its components. At the same time, landscapes from peripheral areas have been used within the interchange area, enriching its environment (Figure 3.22).

Just the name of the Baima Interchange (literally 'white horse interchange') gives one a sense of the historical legends and cultural background here. For this reason, the theme of the interchange design is the white horse's shape, character and spirit. Plants are used to create a 'charming white horse'

(a)

(b)

(c)

(d)

(e)

Figure 3.21   (a–e) The features and charm of the Yanzhi River are expressed in a remarkably lifelike way within the Luojiabian Interchange area.

environment (Figure 3.23). In order to achieve this, the following measures were taken:

(1) Full advantage was made of the hills within the interchange area, arranging landscapes according to local conditions and avoiding mass excavation and filling which might have damaged the local environment.

(2) Different species of plants were planted layer by layer along the contour lines of hills; the altitude difference ensured a clear and three-dimensional view.

(a)

(b)

(c)

(d)

Figure 3.22   (a–d) Groups of big trees and shrubs fully reveal the beauty of the Yanzhi River.

(3)   Evergreen and deciduous plant species with large colour and morphological differences were selected, such as *Magnolia grandiflora*, *Acer palmatum*, *Albizia julibrissin*, *Metasequoia glyptostroboides*, *Taxodium ascendens*, *Liquidambar formosana*, *Sapium sebiferum* and *Lagerstroemia indica*. These plants have different characteristics and complement each other very well.

(4)   Streams within the interchange area were appropriately widened so that they look natural, and species such as weeping willow and peach blossom were planted for people to enjoy.

(5)   In addition to trees, flowering shrubs were meticulously designed to blend into the landscape. The beautiful and fragrant flowers of these plants have provided a pleasant atmosphere for travellers.

(6)   Outside the layer of trees and shrubs, wild flowers have been arranged to improve the local scenery.

The above measures reflect a positive spirit and rural charm; this is the essence of landscape design (Figure 3.24).

The South Liyang Interchange is the nearest interchange to Tianmu Lake, which is a national 4A-level scenic area. It is surrounded by a beautiful environment featuring hills, mountains and trees as well as green bamboo groves and tea plantations on sloping fields. The local topography was adjusted in order to recreate the natural and picturesque scenery of Tianmu Lake during

Figure 3.23   White horse sculpture.

design of the interchange landscape. It is also important to note that the interchange is surrounded by mountains and water (there is a large area of water in the surroundings) and has sloping fields. By excavating ponds, constructing mounds and building slopes, an attractive scene has been created in the interchange area. The area features hillside woodland in addition to wild birds and fish, successfully reflecting the natural features of Tianmu Lake. The following specific measures were employed:

(1)   Ponds were excavated and mounds were constructed (Figure 3.25) within the interchange area to form three hills of different shape and height. The elevation of the hills is higher than the surface of the road. In the middle of the water a small hill was constructed to resemble a hill on an island in Tianmu Lake. Viewed from the periphery of the interchange, these mounds of earth are just like small hills and fit perfectly into the surrounding environment, so much so that they seem

(a)

(b)

(c)

Figure 3.24   Use of existing mounds within the interchange area. Landscapes are made according to local conditions within the interchange area; the layering of plant landscapes is distinct and the open fields become more attractive.

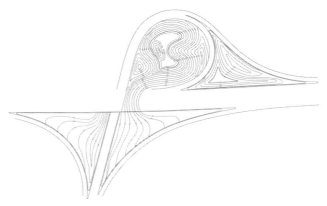

Figure 3.25   Treatment and design of topography.

to be part of the surrounding hills of Tianmu Lake. At the south side of the main route, two triangular areas surrounded by the ramps have been made into gentle slopes. The above design has resulted in a remarkably original lake, woodland and aquatic landscape.

(2) Plant landscapes. The 'small hill area' of Tianmu Lake has been arranged with evergreen trees, such as *Cinnamomum camphora*, *Cedrus deodara*, *Schima superba*, *Michelia* and *Pinus elliottii*, as well as deciduous tree species, such as *Sapium sebiferum*, *Liriodendron chinense*, *Bischofia polycarpa*, *Salix babylonica*, *Camptotheca acuminate*, *Sapindus mukorossi*, *Zelkova schneideriana*, *Ginkgo biloba* and *Magnolia liliflora*. This natural design contributes to a stretch of woodland characterized by a landscape of local subtropical vegetation. In addition, the planting of *Osmanthus fragrans* (four-season variety), *Nandina domestica* and *Hibiscus syriacus* at the edge of the woodland as well as wetland and aquatic plants such as *Iris pseudacorus*, *Pteris multifida* and *Saururus chinensis* at the edge of the water has created an area that features both natural and garden landscapes (Figure 3.26).

The design of the sloping field within the ramp zone south of the main route is abstract and imposing. Various kinds of local bamboo have been shaped into 'swimming fish' patterns of varying height and size that depict fish swimming towards the north of Tianhu Lake. The design of the 'fish' is remarkably lifelike. In order to make the 'fish' seem more animated, flowering shrubs such as *Osmanthus fragrans* (four-season variety), *Camellia sasanqua*, *Rosa chinensis*, *Prunus japonica* and *Buxus megistophylla* have been arranged amongst the bamboo. The resulting landscape features both higher and lower areas, and the differing gradients complement each other well, reflecting the theme of Tianmu Lake (Figures 3.27–3.29). Figures 3.30 and 3.31 show different parts of the interchange.

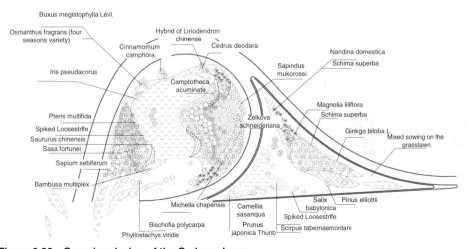

Figure 3.26   Greening design of the Q-shaped area.

Figure 3.27   The landscape within the interchange area is just like an island in Tianmu Lake and fits in well with the surrounding landscape.

Figure 3.28   Various kinds of local bamboo have been made into different fish patterns.

Figure 3.29   Various 'swimming fish' patterns of different height depict fish swimming towards the north of Tianmu Lake.

Figure 3.30   South Liyang Interchange landscape.

Figure 3.31   Small sculptures erected to promote the local scenery and attract tourists.

Yixing Interchange is a double-trumpet interchange, which covers a large area on the outskirts of Yixing. As an agriculturally and industrially advanced region in Jiangsu Province, Yixing abounds in natural resources and has a beautiful natural environment. There are many specialities, such as tea, moso bamboo (*Phyllostachys edulis*) and ceramics. The renowned ceramic teapot industry as one of the area's major industries in connection with tea culture, well-known in China and elsewhere, is a big local industry. By reflecting in the theme the tea and ceramics culture, the landscape design of this interchange introduces the local historical, cultural and folk customs to visitors, so that visitors passing by and entering the urban area may form an initial impression of the city. For those visitors who are already familiar with the city, the design of the interchange can deepen their knowledge of the surroundings.

The undulating hills next to the Yixing Interchange are surrounded by lush green vegetation, including China fir, bamboo groves, tea plantations and mulberry fields. The city's high pagoda is testament to its rich history. Incorporating this view into the interchange has made the landscape even more breathtaking. As this interchange is situated in the suburbs, it has been designed to fit into the green city landscape. In terms of greening design, vegetation patterns have been used in combination with the natural surroundings to reflect local characteristics. In order to achieve this, tea plants have been shaped into teapot patterns on the slopes on both sides of the central ramp on the north side of the north interchange. Rows of tea plants resembling a tea plantation have also been arranged along the topographical contour lines within the south interchange area. In other parts of the interchange area, various kinds of bamboo, evergreen and deciduous trees and shrubs have been combined naturally, meaning that the interchange blends perfectly into the surroundings. Building a pagoda has made the landscape even more stunning. The patterns within the interchange area are striking and distinct, leaving visitors with a lasting impression (Figure 3.32).

Various natural landscapes created to integrate the interchange into the surrounding environment

(1) Woods. Groups of varying numbers of large and small evergreen and deciduous trees have been planted within the interchange area. These groups of trees stand either alone or with others to form a striking woodland landscape. These landscapes consist of different tree species and saplings of various sizes, forming seemingly natural tree canopies of differing heights. In some cases, woods have been arranged beside the guardrail, making the motorway seem as though it is passing through forest. In order to make sure that the tree landscape is as natural as possible, natural vegetation has been simulated within the interchange area, reproducing a beautiful natural landscape (Figure 3.33).

(2) Grass field. A natural landscape is reproduced within the interchange area to reveal a grassland landscape. For example, the Guizhuang

Figure 3.32 (a–g) Bamboo groves, tea plantations, mulberry fields, tourism advertisements and a pagoda have made the landscape around the Yixing Interchange even more stunning.

(a)

(b)

Figure 3.33 (a, b) Tree landscape within the Guizhuang Interchange area.

Interchange covers an area in excess of 100 000 square metres, of which approximately 30 000 square metres contains wild flowers and grass. This landscape accentuates the natural, beautiful and harmonious atmosphere of the tree and wetland environment, satisfying the demands of people today for a scenic environment. The main species include *Iris pseudacorus*, Spiked Loosestrlfe, *Orychophraqmus violaceus*, *Mirabilis jalapa* Linn., *Indocalamus tessellatus*, *Arundo donax* (var. *versicolor*), *Polygonum orientale* Linn. and *Saururus chinensis*. These wild flowers and grass are distributed around the trees in an arc or curve arrangement to add extra layers and colour to the trees. The result is a natural and wild grass landscape (Figures 3.34 and 3.35).

(3) Large grassland landscape. In modern garden design, grass accounts for a larger proportion of land than in traditional garden design.

Figure 3.34 *Phragmites communis* **on the grass field becomes yellow and the grassland is green.**

Figure 3.35   **Grassland interspersed with flowers at the edge of the trees.**

During design of the interchange, full attention has been paid to people's environmental requirements by ensuring a large flat area of land within the interchange zone. In some interchange areas, grassland covers nearly half of the total area. In order to keep the grassland green all year round, warm-season and cold-season grass species are combined to form a composite grassland species. The green grass is planted amongst trees or on its own, forming a large area of grassland. Planting the grass in woodland amongst dense trees and shrubs makes the landscape seem even more spacious. Moreover, a combination of wild flowers has been planted on the wide expanse of grassland to create a beautiful landscape of grassland interspersed with colourful flowers (Figure 3.36).

(4)   Flowering shrubs. Large-area shrubs with colourful leaves and flowers are combined at the edge of woodland and near residential areas to add layers and colour to the landscape.

Figure 3.36 Broad grassland distributed in the underwood.

(5) Bamboo groves and tea plantations. Bamboo groves and tea plantations can be found all along the route of the Nanjing-Hangzhou Motorway. For the purpose of blending into the surrounding environment, clusters, ridges and plots of bamboo and tea plants are arranged within the interchange area. The unique sight of bamboo and tea plants reflects the environmental characteristics of hilly and mountainous regions south of the Yangtze River, leaving people with a lasting impression (Figure 3.37).

A wetland draining system is used to replace the conventional draining ditch

All conventional interchange draining systems have been replaced by wetland, which is a great innovation in motorway interchange draining systems as well as a pioneering undertaking in landscape design.

Figure 3.37 Bamboo and tea trees within the interchange area.

The marshes and wetlands are built near ponds, low-lying areas, drainage ways and outfalls in various interchange areas. For example, *Phragmites communis* is planted in strips in the bottom-land recess areas. In addition to the *Phragmites communis* that is planted beside the ponds and outfalls, other aquatic and wetlands plants are introduced, such as *Iris pseudacorus* and *Typha angustifolia*, thereby accentuating the wet land scenery in the interchange area. Wetlands are one of the most important and common natural landscapes in nature and are very biodiverse. Wetlands are known as the earth's 'kidneys' because they have a strong circulatory function for water and elements. Humans cannot live without them, and neither can many organisms. As they can store, retain and purify water, wetlands can moisten the land and clean muddy water and polluted water. Wetlands not only have the ability to protect the environment, they are also an essential part of nature.

A natural and harmonious wetland landscape restores the damage caused by artificial engineering and represents progress in engineering design and construction. All the interchange areas on the Nanjing-Hangzhou Motorway use wetlands for their water treatment systems and have set an example for the draining engineering of future motorways (Figures 3.38 – 3.40).

Attach importance to the reshaping of the topography

During the environmental landscape design of the Nanjing-Hangzhou Motorway, great importance has been attached to reshaping and reconditioning of the topography, such as reducing the gradient of the subgrade slopes; some are lowered to the same level as the side ditch stage. By lowering the slopes, mortar rubble is not required for slope paving, which means that the green area along the motorway can be increased, thus creating conditions for various natural landscapes. The difference in height between the slope and the subgrade is small, providing a good viewing angle for motorway users, leaving them with a wider field of view and sense of security.

On major landscaping sections such as interchanges and service areas, the local topography is reconditioned, ensuring that the landscape under construction can blend into the natural environment. This also creates favourable

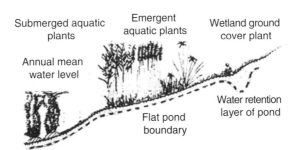

Figure 3.38 Constructed wetlands and wetland plant distribution sequence.

Figure 3.39   Wetland landscape within the Guizhuang Interchange area.

Figure 3.40   Poetic and picturesque scenery within the Shangxing Interchange area.

conditions for the layout and planting of green plants. The excavation of ponds and the construction of artificial hills make for a natural environment and the undulating landscape blends well into the surroundings. The planting of vegetation should be as natural as possible and efforts should be made to make the surroundings at one with nature. For example, bamboo and tea plantations on both sides of the subgrade have been integrated into those outside the motorway boundaries, resulting in a coordinated green landscape along the entire route.

At the east Lishui Interchange, narrow ramps cover a large area of land. It is also difficult to enter the interchange and the draining system is complicated, meaning that the land is unsuitable for crops. For this reason, the builders redetermined its purpose and function and decided to build the motorway

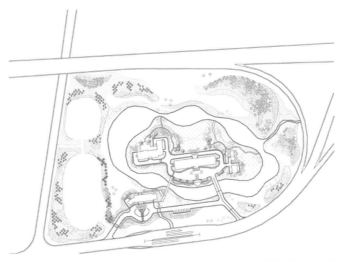

Figure 3.41   Floor plan of the traffic management centre within the east Lishui Interchange.

management centre here (Figure 3.41). In terms of landscape planning and reconditioning, the following techniques were used:

(1) Water layout. The water system was dredged and water landscapes created. In order to create a landscape the originally separated ponds and streams were linked together and then widened into a ring-like curve according to their shape; overflow dams were designed according to drainage height so as to store water. Increasing the water surface levels and depth of field has made the area feel more spacious. These waterways are wide in some places and narrow at others, generally used to store water but also to drain away water during rain. The natural flow of water creates a sense of tranquility and warmth.

(2) Artificial islands. The originally disorderly borrow pits have been reconditioned, and the earth excavated from widening the water channel has been placed in the middle to form artificial hills, the tops of which have been flattened for the construction of the motorway management centre. The buildings stand on the resulting island, forming a quiet and gentle waterscape.

(3) Waterweeds. Waterweeds, such as *Phragmites communis*, *Iris pseudacorus* and *Caldaria parnassifolias* have been planted beside the bank to accentuate the natural features of the water landscape.

(4) Flowering trees. Trees, shrubs and ground cover plants have been arranged around the buildings and at both sides of the ramp to form a combined landscape. For this purpose, many species of green plants were selected, including more than 40 species of trees. Most of the trees are native species, such as *Castanopsis sclerophylla*, *Liquidambar*

formosana, Pterocarya stenoptera, Cinnamomum camphora, Koelreuteria paniculata, Metasequoia glyptostroboides, Salix babylonica, Magnolia grandiflora and *Osmanthus fragrans.*

(5)   Plant grass. Combinations of wild flowers including *Papaver rhoeas, Echinacea purpurea, Delphinium grandiflorum, Cosmos bipinnatus* and *Rudbeckia laciniata* have been arranged on the grassland near the open water bank. These colourful grass flowers add to the beauty of the green grassland.

(6)   Car parks. In the area between the west side of the toll station, management centre and the local road, car parks have been built, surrounded by a belt of large trees that provides safe coverage for transport vehicles.

(a)

(b)

(c)

(d)

(e)

Figure 3.42   (a–e) Landscape surrounding the traffic management centre.

From the main route, drivers can see attractive yellow-walled and blue-tiled buildings of both Western and Chinese design on the distant island. Both tall and short flowering shrubs, deep-green grass and flowers can be seen amongst the buildings. At a closer distance all kinds of flowering shrubs with colourful flowers can be seen dotted around the grassland. Weeping willows and blooming peach blossom make for a very attractive landscape (Figure 3.42). If travellers look at the landscape on the ramp leading to the toll station, they can see large plots of tall evergreen trees that momentarily obscure the island and the buildings, in addition to being able to see most of the features that can be found from the main route. The attractive landscape leaves people with a lasting impression.

Situated north of the picturesque and verdant Dashishan Mountain in Liyang, the west Liyang Interchange is a single-trumpet interchange, with its high point in the southeast and low point in the northwest. The motorway passes through two neighbouring villages, and the huge interchange is located between such two villages, one of which lies to its northwest and the other to its southeast (Figure 3.43). The motorway structures rise abruptly above the flat ground below, meaning that the original natural landscape has been

(a)

(b)

Figure 3.43   (a, b) The interchange is situated between two villages.

subject to great change and the lives and farming of local residents has also been affected. The motorway has also had a significant visual impact on residents and the many pollutants resulting from the motorway have directly affected their health and the quality of the surrounding environment. In order to mitigate these negative impacts, the interchange landscape design has focused on creating landscapes that will improve the appearance of the surroundings and prevent pollution by planting vegetation and building protective forest belts. The specific measures adopted are as follows:

(1)   Establish the creative theme of landscape design. Artificial mounds within the interchange area will become small hills of the Dashishan Mountain range. By covering it with forest, it will become part of the natural landscape (Figure 3.44).

(2)   Recondition the topography. In order to proceed with the themed design, the topography within the interchange area should be renovated and reconditioned (Figure 3.45). In order to achieve this, a few mounds higher than the ramp should be built with earth from excavating ponds. The concrete measures are as follows:

(a)   Within the Q-shaped area encircled by ramps B and E, dig a ditch at the edge, borrow earth and build an 18-m-high mound that is 4 m higher than the ground level of the southeast neighbouring village.

Figure 3.44   Creative design of green landscape.

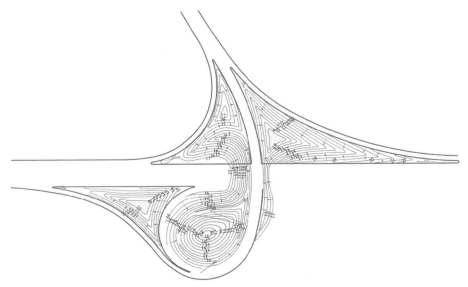

Figure 3.45   Floor plan of reconditioned topography within the interchange area.

Looking north and west from the village, the ramp cannot be seen; only the mound can be seen. For the shape of the mound and its relationship to the motorway and ramps, see Figure 3.46, sections 8, 9, 10 and 11.

(b)   Dig a pond and borrow earth in the middle of the triangular land area enclosed by ramp A, then build a mound with a perimeter of 17 m and 7 m higher than the ground level outside the ramp. It looks like another hill from a distance.

(c)   Dig a pond in the middle of the triangular area encircled by ramp D and build a mound 6 m higher than the ground level outside the ramp. See Figure 3.46, sections 1, 2, 3 and 4. It also looks like a hill from a distance.

(d)   Create an undulating landform within the triangular area surrounded by ramp C. It should be 8 m higher than the ground level outside the ramp. See Figure 3.46, sections 5, 6 and 7.

The above reconditioning results in a more diverse topography due to the addition of mounds, ponds and slopes, and this diverse topography is also conducive to the arrangement and landscaping of plants. When facing one village from the other one, several high and low mounds can be seen rather than the high subgrade of the motorway. These man-made mounds fit well into the rural environment and greatly alleviate the visual effect on the surrounding residents. In terms of engineering, the excavation and heaping of earth on site is an effective means to deal with earthworks. Filling is carried out in a balanced manner, saving investment (Figure 3.47).

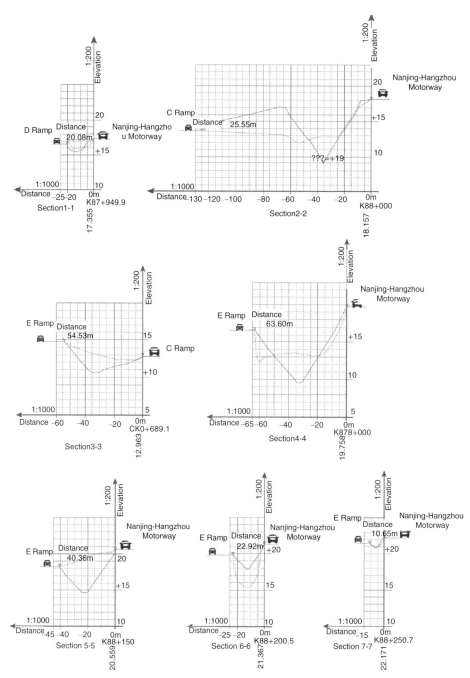

Figure 3.46 Perspective view of topographical treatment.

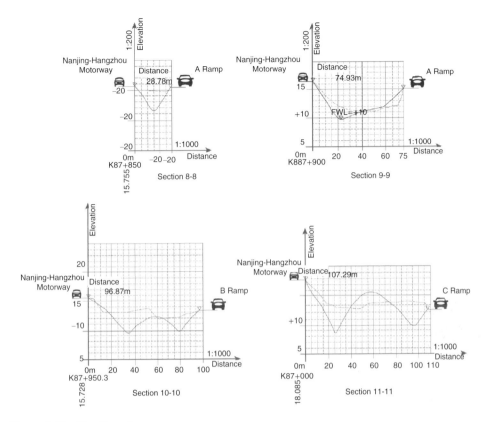

Figure 3.46 *(Continued)*

(3) Create garden landscapes and protection forest belts. After topographical treatment, create a themed landscape, imitate the natural vegetation of the Dashishan Mountain, and densely plant woods and shrubs on the mounds. The aim is in essence to copy and reproduce the beautiful scenery of the Dashishan Mountain. In order to create these scenes, broad-leaved evergreen tree-based species, such as *Cinnamomum camphora*, *Cryptomeria japonica* (var. *sinensis*), *Elaeocarpus decipiens*, *Ilex chinensis*, *Eriobotrya japonica* and *Magnolia grandiflora* are arranged on the mound in a natural manner. Evergreen shrub species, focusing on native bamboo and *Camellia oleifera*, such as *Mahonia fortune*, *Nerium oleander*, *Nandina domestica*, *Pyracantha fortuneana* and *Yucca gloriosa* are also planted. Meanwhile, some deciduous trees and shrubs are also selected, such as *Bischofia polycarpa*, *Celtis sinensis*, *Liquidambar formosana*, *Sapindus mukorossi*, *Sapium sebiferum* and *Acer palmatum*. The groups of plants composed of these key tree species have relatively similar morphological characteristics to the local natural vegetation and fit well into the neighbouring environment, making for a beautiful landscape (Figure 3.48). *Edgeworthia chrysantha* and *Nandina domestica*

Figure 3.47   By digging ponds and building mounds within the Q-shaped area of the interchange, a natural landscape is created.

(a)

(b)

Figure 3.48   (a, b) After greening, the rural scenery is natural and beautiful.

are arranged over a large area in the area encircled by the ramps. This is in line with ecological principles, because the former is a heliophilous deciduous plant, while the latter is an evergreen plant which grows well in an environment with partial shade. The former blossoms before its leaves appear; its flowers are unique and beautiful, and its branches give shade to the *Nandina domestica*. In late autumn the *Edgeworthia chrysantha* sheds its leaves, leaving sufficient sunlight for the *Nandina domestica*,

(a)

(b)

Figure 3.49   Trees planted on several mounds higher than the ramps obscure the unattractive motorway landscape.

whose leaves turn from green to red. The branches of the *Nandina domestica* become dripping in red fruit, creating an attractive scene for visitors in winter.

In addition, a wide protective belt consisting of trees and shrubs has been built next to all the ramps in the interchange area. This protects the residents against noise, dust and exhaust pollution. The interchange has become obscured from view due to the dense branches of the forest belt, turning the place into a popular scenic spot (Figures 3.49–3.51).

Figure 3.50 Wide protective tree and shrub belts have been built next to the ramps.

Figure 3.51 The west Liyang Interchange fits into the surrounding environment.

4 Environmental Landscape Design of the Central Reservation

4.1 Definition

The area separating the two-direction carriageways of the motorway is known as the central reservation (Figures 4.1 and 4.2). The central reservation varies in width. Narrow ones are just tens of centimetres. There are many forms of central reservations, such as median barriers, cement blocks, anti-glare shields and boundary markers (Figures 4.3–4.5). If the central reservation is more than 1.5 m in width a green belt is often used. In Jiangsu Province, the central reservation of motorways is generally about 3 m wide, for which green plants are used. In developed countries there is generally no central reservation on the motorway. If there is a central reservation, it is wide and generally made of vegetation. In some cases, the overtaking lane and the central reservation are built together, so it is much wider (Figures 4.6 and 4.7). In Europe and the USA, the plants in the central reservation are left to grow naturally according to local conditions; even overgrown wild grass is not pruned. It therefore fits in well with the natural landscape on both sides. The type of each green central reservation is also different (Figure 4.8). Research has proven that if the central reservation is more than 12 m wide, vehicle lights from the other side of the motorway do not affect drivers even without the help of an anti-glare barrier. This is an added benefit when constructing a central reservation.

4.2 Functions

If plants are used as boundary markers for the central reservation, the design of the green landscape is very important, because it will be seen by motorway users as they drive past. Research shows that as vehicles speed up, the driver's

The Environment and Landscape in Motorway Design, First Edition.
Qian Guochao, Tang Shuyu, Zhao Min and Jing Chun.
© 2014 China Communications Press. Published 2014 by John Wiley & Sons, Ltd.

Figure 4.1   Central reservation.

Figure 4.2   Central reservation.

Figure 4.3   Separation with an anti-glare shield.

Figure 4.4   Separation with a separation net.

Figure 4.5   Separation with soft material.

Figure 4.6   Central reservation from outside China.

Figure 4.7   Central reservation from outside China.

Figure 4.8   Greening of a central reservation on a motorway from outside China.

attention is attracted towards the lane and they focus only a small range; their field of vision therefore becomes very narrow and eventually results in so-called 'tunnel vision'. This means that the central reservation accounts for an increasingly larger proportion of the driver's field of vision, whilst the landscape on either side of the motorway accounts for an increasingly smaller proportion. The attractiveness of the green central reservation therefore has a direct impact on the driver's vision. Considering the safety of motorway users, the primary function of the central reservation should be anti-glare, namely preventing vehicle lights from causing glare to drivers travelling in the opposite direction at nighttime, which can result in traffic accidents and a sense of danger when passing other vehicles. The secondary function of the central reservation should be greening. Green plants in the central reservation attract drivers' attention due to the change in their colour and shape. This guides traffic and lessens driver fatigue, especially mental and visual fatigue. Green central reservations can also regulate drivers' mood and efficiency. Thirdly, landscape design of the central reservation should focus on improving the appearance of the environment, creating an attractive and beautiful scene for motorway users as well as building a safe and comfortable traffic environment. Green central reservations such as these can generate interest in the motorway and turn it into a tourist attraction (Figures 4.9 – 4.11).

Figure 4.9 **The white flowers of** *Yucca gloriosa* **make the central reservation more attractive.**

Figure 4.10 *Rosa chinensis* **beautifully adorns the central reservation.**

Figure 4.11   Camellias add attractive colours to the central reservation during the winter.

4.3 Landscape design

Design principles

(1)   Satisfy traffic demands and ensure good anti-glare performance.
(2)   Plant evergreen plants and anti-glare trees that have a stable morphology anti-glare effect.
(3)   Combine coloured-leaf shrubs with flowering shrubs to achieve rich colours, and a layered and attractive view.
(4)   Choose hardy and stress-resilient plants that are resistant to the adverse motorway environment.

Design space

The basic form of the central reservation is about 3 m wide. If a figure of 3 m is used, the planar space for plants is only 2.7 m in width after deducting 30 cm (15 cm+15cm) for the thickness of the curb; if plants are arranged between median barriers (the distance between the median barrier and the curb is 50 cm, and 1 m in total), the spatial distance is only 1.5–1.7 m. Looking at a cross section of the central reservation, the vertical height between the curb and the median barrier is only 25–45 cm and the bottom is slanted (Figure 4.12). This means that within such area only undershrubs and perennial flowers with undeveloped root systems can be planted; such an area is not suitable for planting trees. However, the vertical height of the 1.5–1.7 m median barrier is 55–95 cm. As there is a plastic discharge pipe in the centre resulting in a deep centre and shallow edges, shrubs or shallow-rooted trees may be planted, preferably on both sides.

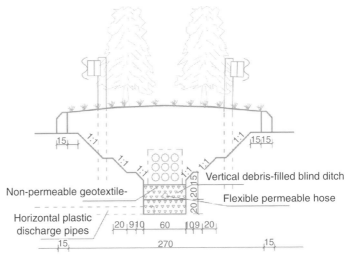

Figure 4.12   Cross section of the central reservation (unit: mm).

Design requirements

The primary function of greening the central reservation should be anti-glare. The anti-glare light shielding angle should be in the range of 8–15°. Full consideration should be given to negative factors relating to the central reservation, such as small space, barren soil, low humidity, heavy pollution and maintenance difficulties. The arrangement of plants should be carefully researched, as should the selection of tree species, flowers and grass species. As for the landscape design of the central reservation, vehicle speed should be taken into full account; the landscape design of particular sections should vary based on vehicle speed. As vehicle speed increases, the dimensions of the landscape should increase. At a speed of 120 km/h, a car travels 2 km in 1 min, so in terms of the driver's vision, 2 km of scenery passes by in just 1 min. Research has shown that it takes 5 s to see a scene clearly. The distance that a car travels within these 5 s is known as the unit length of a scene. When a car travels at a speed of 120 km/h, the unit length of a scene is 166.5 m and at 140 km/h it is 194.5 m. This means that the only way to leave drivers and passengers with a lasting impression is to increase the dimension of the landscape. The length of plant landscapes, such as trees and coloured vegetation belts should be in the range of 200–300 m. The landscape should change at intervals of 6–10 km. Whilst ensuring unity there should also be variations and not too much repetition. The central reservation should be green all year round with flowers and trees. The landscape should be neat and orderly, changing as vehicles move along the motorway.

Anti-glare design

Planting of anti-glare species: Central reservations can easily provide an anti-glare effect. As long as trees and shrubs are planted into green rows in the central reservation, there will be an anti-glare effect. However, this form of anti-glare greenery is both unattractive and expensive. On the other hand, if plants are arranged at random, they look sparse and natural as well as being cheap to employ but they cannot provide a satisfactory anti-glare effect. Practice and research has shown that using plants in the central reservation for preventing glare depends on the features and arrangement of the plant species selected. The anti-glare effect is related to the morphology, diameter and density of the anti-glare plant species. The light shielding angle of the anti-glare trees and their combination of vegetation must be more than $7-15°$ before they have an anti-glare effect.

The plant anti-glare test formula is:

$$L = M/\tan \alpha$$

L = upper limit of distance between plants (m)
M = diameter of the plant 0.9 m from the ground
α = diffusion angle of the car headlight, usually 8°.

According to the formula, when M is 0.5 m, the light shielding angle is $8-15°$, indicating an anti-glare function.

In addition, headlights are installed at different heights according to the size of vehicle. In order to prevent glare, the height of plants shall be controlled

Figure 4.13 *Sabina komarovii.*

between 1.5 m and 1.7 m. If they are too tall then they will cast shadows onto the surface of the road when there is sunlight. The taller they are, the longer the shadow. Shadows, which are also a kind of glare, interfere with drivers' vision and cause fatigue.

In order to achieve an optimal anti-glare effect, there are relatively stringent requirements for anti-glare trees. The specific requirements are as follows:

(1) Evergreen, short leaf-changing period and minimal loss of leaves.
(2) Column-shaped with substantial branches and leaves, dense, conducive to pruning and easy to shape.
(3) Resilient to drought, barren soil and pollution.
(4) Slow-growing and long lifespan.
(5) Do not release unpleasant odours or large quantities of pollen.

Years of practice in Jiangsu Province have proven that the tree with the best anti-glare effect is *Sabina komarovii* (Figure 4.13), followed by *Juniperus chinensis* (Figure 4.14), *Viburnum odoratissimum*, *Photinia*

Figure 4.14 *Juniperus chinensis.*

serrulata (Figure 4.15), *Buxus megistophylla* (Figure 4.16), *Ligustrum quihoui* (Figure 4.17) and *Pittosporum tobira* (Figure 4.18).

On the Nanjing-Hangzhou Motorway (Phase I), the anti-glare plant species chosen for the central reservation were *Osmanthus fragrans* and *Trachycarpus fortunei*. With hindsight, both are inferior to *Sabina komarovii* in terms of their anti-glare effect. *Osmanthus fragrans* suffers from substantial leaf loss and is not very resilient. In addition, its branches and leaves are sparse, meaning that the anti-glare effect is unsatisfactory. This is even more so the case with *Trachycarpus fortunei*. Its survival rate after transplanting is low and it does not recover easily. It has a small trunk and does not have many leaves, therefore

Figure 4.15 *Photinia serrulata.*

Figure 4.16 *Buxus megistophylla.*

Figure 4.17 *Ligustrum quihoui.*

Figure 4.18 *Pittosporum tobira.*

it cannot effectively shield light and prevent glare. These problems are related to not combining the trees with other plants during the design of the central reservation as well as poor construction quality and management.

Combination of flowering shrubs: During the design of central reservation, in addition to choosing optimal anti-glare plant species and determining the distance between each plant, attention should also be paid to an appropriate combination of other flowering shrubs and ground cover plants with the anti-glare species. This is necessary to create a green and attractive environment and avoid a dull and monotonous landscape. Meanwhile, it is essential to ensure that there is colour contrast and that the plants complement each other well. It is also important to note the effect that different seasons and

(a) Cercis chinensis

(b) Hibiscus syriacus

(c) Spiraea prunifolia

(d) Camellia japonica

(e) Rosa multiflora

(f) Kerria japonica

(g) Spiraea cantoniensis

(h) Prunus japonica Thunb

Figure 4.19   (a–h) Some combined flowering shrubs.

colours have on motorway users. For example, in summer, it is best to choose flowers with a lighter colour; in autumn and winter, it is better to select maple leaf trees and species with red flowers; and in spring, it is advisable to choose species with yellow and pink flowers. Flowers and ground cover plants commonly used in combination include:

(1) Shrubs: *Lagerstroemia indica, Hibiscus syriacus, Punica granatum, Cercis chinensis, Malus halliana, Prunus persica, Spiraea cantoniensis, Osmanthus fragrans, Pittosporum tobira, Prunus cerasifera* Pissardii, *Cornus alba, Buxus megistophylla, Platycladus orientalis, Gardenia jasminoides, Ligustrum quihoui, Ligustrum ovalifolium* Vicaryi, *Forsythia viridissima, Pyracantha fortuneana, Loropetalum chinense* (var. *rubrum*), *Berberis thunbergii, Camellia sasanqua, Camellia japonica, Camellia oleifera, Rhododendron simsii, Hypericum monogynum* and *Yucca gloriosa* (Figures 4.19–4.21).

(2) Ground cover plants: *Cynodorz ctylorz, Trifolium repens, Dichondra repens, Zephyranthes candida, Oxalis rubra, Ophiopogon japonicas, Ophiopogon japonicas, Setcreasea purpurea, Cynodon dactylon, Rudbeckia hirta, Fatsia japonica, Reineckia carnea* and *Rosa chinesis.*

(3) Flowers: *Canna indica, Rosa* cultivars, *Rosa chinensis* Jacq., *Rosa multiflora, Kerria japonica, Lycoris radiate, Iris tectorum* and *Orychophragmus violaceus* (Figures 4.22 and 4.23).

Design of normal section

The normal section is the main part of the central reservation, accounting for about two-thirds of the total length of the motorway. The purpose of its design is to create an anti-glare effect. Once this has been achieved then only a few

(a) Prunus persica   (b) Malus halliana

Figure 4.20   (a, b) Some combined flowering shrubs.

(a) Forsythia viridissima

(b) Lorpetalum chinense

(c) Lagerstroemia indica

(d) Kerria japonica

(e) Platycladus orientalis

(f) Camellia sasanqua

(g) Pyracantha fortuneana

(h) Camellia japonica

Figure 4.21   (a–h) Some combined flowering shrubs.

(a) Orychophragmus violaceus

(b) Zephyranthes candida

(c) Papaver rhoeas

(d) Lycoris aurea

(e) Oxalis rubra

(f) Canna

Figure 4.22   (a–f) Some combined ground cover plants and flowers.

measures need to be taken to improve its appearance. This section is usually straightforward, simple and spacious. The specific design is as follows:

(1)   Size of anti-glare plant species. Using *Sabina komarovii* as an example: its cut stem is 1.6 m high, its diameter at the height of 90 cm from the ground is 50 cm, and the cut top crown diameter is 15 cm.

(2)   Planting form and spacing. The *Sabina komarovii* is planted in a single row at intervals of 2 m; if it is planted in double rows, the spacing is not more than 4 m (Figure 4.24).

(a) Dichondra repens

(b) Festuca arundinacean

(c) Vinca major

(d) Cynodon dactylon

(e) Trifolium repens

Figure 4.23   (a–e) Several common ground cover plants.

(3)   Arrangement of shrubs. Coloured-leaf, evergreen shrubs and spherical plants can be planted between and at the sides of the main anti-glare tree species (*Sabina komarovii*) in order to form a simple combination of different heights and colours (Figure 4.25).

(4)   Ground cover plants. Some species that cover the ground should be selected, such as *Trifolium repens*, Bermuda grass, *Ophiopogon japonicus* and *Zephyranthes candida*.

(a)   (b)

Figure 4.24   (a, b) Simplest design of the normal section.

Figure 4.25 *Ligustrum ovalifolium* **Vicaryi between the median barrier and curbs.**

Accentuating section (section to improve the landscape)

Such a section is designed to improve those landscapes in the normal section with a length of approximately 10–20 km, as well as in sections adjacent to service areas, toll stations, residential areas, tunnels, bridges and urban areas. The design aims to improve the appearance of the normal section and bring the landscape to life. The accentuating section is designed by adjusting the arrangement of anti-glare trees and planting numerous coloured-leaf shrubs and perennial flowers to create large, bright and smoothly coloured blocks and shapes (Figures 4.26–4.28). The normal and accentuating sections appear alternately along the motorway. Generally speaking, 1–2 km of accentuating section will occur after 10–20 km of normal section.

By looking at the design drawings it can be seen that the arrangement of the primary anti-glare tree (*Sabina komarovii*) varies significantly. As shown in Figure 4.26, it is arranged on one side of the median barrier; in Figure 4.27, a section of *Buxus megistophylla* is separated by a section of *Sabina komarovii*;

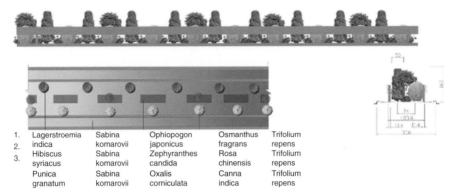

1.	Lagerstroemia indica	Sabina komarovii	Ophiopogon japonicus	Osmanthus fragrans	Trifolium repens
2.	Hibiscus syriacus	Sabina komarovii	Zephyranthes candida	Rosa chinensis	Trifolium repens
3.	Punica granatum	Sabina komarovii	Oxalis corniculata	Canna indica	Trifolium repens

Figure 4.26   Design I of the accentuating section.

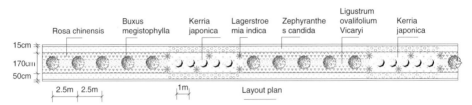

Figure 4.27   Design II of the accentuating section.

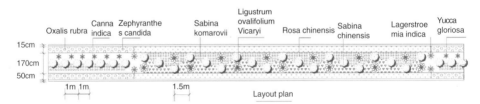

Figure 4.28   Design III of the accentuating section.

in Figure 4.28, rows of *Sabina komarovii* (3 plants) are arranged like a shutter blind in the middle of rows with only one plant. In these sections, trees are also combined with many coloured-leaf and flowering shrubs to form large coloured blocks and clusters. This results in an effective colourful environment. Figures 4.29–4.31 show the three effectsof the designed accentuating section of the motorway.

Design of the safety island end

The safety island end refers to the opening of the independent unit of the central reservation. Short plants are arranged first before gradually increasing the height of the plants until they are the same height as the anti-glare trees in the

Figure 4.29   Design effect I of the accentuating section.

Figure 4.30   Design effect II of the accentuating section.

Figure 4.31   Design effect III of the accentuating section.

normal section. It is usually 5 – 15 m in length, and anti-glare trees are planted densely at an interval of less than 50 cm before being pruned to a slope with a gradient of 1:10. Evergreen and coloured-leaf undershrubs as well as perennial flowers are planted on both sides of the primary tree species and arranged into patterns or coloured blocks (Figures 4.32 – 4.34).

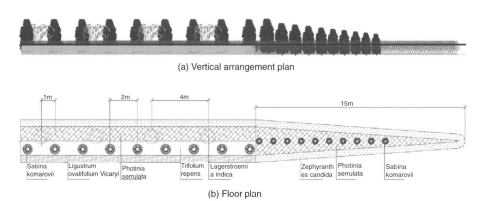

(a) Vertical arrangement plan

(b) Floor plan

Figure 4.32   (a, b) Greening design of the safety island end.

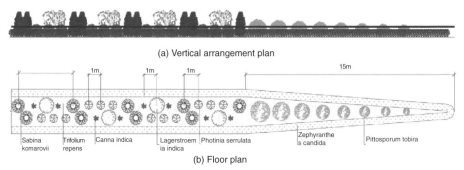

(a) Vertical arrangement plan

(b) Floor plan

Figure 4.33   Greening design of the end of the safety island.

Figure 4.34   Greening effect of the end of the safety island.

4.4   Design pattern

Regular types

> For a central reservation 3 m or below in width, there is little space for plants, and it is necessary to plan the layout well. In general, there are the following several types.

Single row type

The major anti-glare tree species are planted in the middle or at one side of the central reservation according to calculations to determine the interval between them. If the trees are planted in the middle, flowering or coloured-leaf under-shrubs are arranged at intervals between the major anti-glare trees; flowering and coloured-leaf shrubs are used to form strips or blocks and placed on both sides of the trees. Herbaceous flowers can be planted in the middle of the shrubs. In Figure 4.35, the major anti-glare tree *Sabina komarovii* is planted in the middle of the central reservation. *Lagerstroemia indica* is arranged in between the trees, and *Ligustrum ovalifolium* Vicaryi or *Berberis thunbergii* are arranged on both sides at regular intervals. *Rosa chinensis* Jacq. is planted in between with the ground cover plant *Trifolium repens*.

If the major anti-glare tree is planted on one side, it is often designed as a section on the left followed by another section on the right. Each section is 100 – 200 m in length, and may be longer under special circumstances. This form of arrangement is called a 'one-sided approach'. The anti-glare effect of this approach is the same as placing the trees in the middle, but its biggest advantage is that the planting of *Sabina komarovii* on one side of the median barrier frees up space on the other side, where flowering shrubs may be planted to improve the central reservation landscape. This type of landscape was first used along the Wuxi – Jiangyin Motorway to good effect. It is now employed along several motorways in Jiangsu Province.

Figure 4.35   Greening effect of single row type.

Figure 4.36 shows a design for the 'one-sided approach'. The anti-glare tree is arranged on one side adjoining the median barrier. Flowering undershrubs including *Lagerstroemia indica* and *Distylium racemosum* are planted in the spacious area on the other side. *Canna indica* and *Rosa chinensis* are planted next to the undershrubs. *Ligustrum ovalifolium* Vicaryi and *Berberis thunbergii* are deployed at intervals along the side of the curb. *Rosa chinensis* blossoms from May to November every year. *Canna indica* flowers from June to November, and *Lagerstroemia indica* blooms from July to September. Golden *Ligustrum ovalifolium* Vicaryi and red *Berberis thunbergii* keep their colors even longer. Such a colourful and dynamic combination within the green belt provides motorway users with an attractive environment (Figure 4.37).

Figure 4.36   Design I of the 'one-sided approach'.

Figure 4.37   Design II of the 'one-sided approach'.

Double row type

This refers to two rows of anti-glare trees forming a green belt inside the median barrier of the central reservation. Compared with a single row of trees, a double row has the same anti-glare effect, the only difference being a different aesthetic result. The anti-glare trees should be arranged at intervals of 4–6 m within each row and both rows should be arranged in an alternate manner. The transverse row spacing of the anti-glare trees should be 60–70 cm and the longitudinal row spacing should be 2–3 m. Between the major anti-glare trees, flowering shrubs may be arranged to vary the colour and increase the diversity of the landscape. Compared with a single row of trees, the green belt resulting from a double row of trees is more substantial, colourful and varied. Figures 4.38 and 4.39 show the plan and elevation arrangement of the double row of trees.

Figures 4.40–4.43 show green belts consisting of double rows of trees in the central reservation of the motorway in Jiangsu Province.

Alternating spherical and column-shaped plants

When planting a single row of anti-glare trees, spherical plants are arranged among the primary trees to form an alternating spherical and column-shaped pattern. Alternatively, 3–5 anti-glare trees are planted after 5–7 spherical plants, forming a spherical-oriented pattern. Regardless of the pattern used, the spherical plant selected must generally be over 1.2 m in height and about 1 m in crown diameter in order to exert an anti-glare effect. The pattern should be of alternating height to achieve a favourable landscaping effect (Figure 4.44).

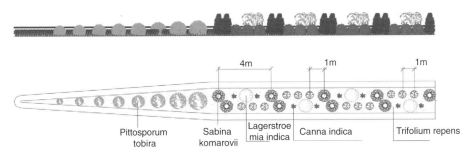

Figure 4.38   Arrangement plan providing a bird's-eye view of the double row of trees.

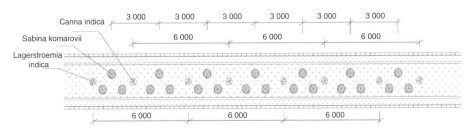

Figure 4.39   Plane sketch of the double row arrangement (unit: mm).

Figure 4.40   Green belts consisting of double rows of trees in the central reservation of the motorway in Jiangsu Province.

Figure 4.41   Green belts consisting of double rows of trees in the central reservation of the motorway in Jiangsu Province.

Figure 4.42   Green belts consisting of double rows of trees in the central reservation of the motorway in Jiangsu Province.

Figure 4.43   Green belts consisting of double rows of trees in the central reservation of the motorway in Jiangsu Province.

(a)

(b)

Figure 4.44   (a, b) A spherical and column-shaped landscape.

Hedgerow type

Anti-glare species with a small crown can be planted in a rectangular volume (30–50 m in length by 0.7 m in width by 1.6–1.7 m in height), in which 3–5 standard anti-glare trees or spherical plants are planted to form a rectangular hedgerow, known as the hedgerow type. This type can be found on the Suzhou section and Zhenjiang sections of the Nanjing–Shanghai Motorway, and along the Suzhou–Jiaxing–Hangzhou Motorway. Its advantages include a good anti-glare effect and a neat appearance (Figure 4.45) but its disadvantage is that it can become too big after it grows to a width of 1 m.

Shutter blind type (inclined row type)

The major anti-glare trees, with three plants in a row, are arranged at 45° to the main route at intervals of 4 m; flowering shrubs and ground cover plants are planted between such two rows. This kind of combination is varied, neat and orderly. This type is applied in the central reservations of

(a)

(b)

Figure 4.45   (a, b) A hedgerow type landscape.

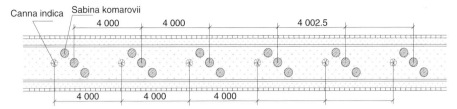

Figure 4.46   Design scheme sketch of the shutter blind type with three plants in a row.

Figure 4.47   Shutter blind type.

the Nanjing–Jingjiang–Yancheng Motorway, the Xuzhou–Suyu Motorway and the Lianyungang–Xuzhou Motorway. In addition to *Sabina komarovii*, *Viburnum odoratissimum* and *Juniperus chinensis* may also be selected as the primary anti-glare trees. *Canna indica* or *Lagerstroemia indica* can be combined with the ground cover plants *Zephyranthes candida* and *Trifolium repens* (Figures 4.46 and 4.47).

Natural type

Central reservations over 5 m in width may be arranged with plants in a natural manner. The wider the central reservation, the more diverse and natural the arrangement of plants can be. The natural arrangement technique does not involve planting trees at random but rather ensuring that ecological and aesthetic requirements are met, as well as satisfying traffic needs. Characteristic plant landscapes are therefore created by carefully imitating natural vegetation and using a combination of trees and shrubs, which should also fit into the surrounding environment. The requirements for this design are quite demanding, so the plant species should be selected appropriately and their combination and arrangement should be rational and aesthetic. This calls for a combination of science and art. This type of design requires micro-topographical treatment in order to make various kinds of landscapes. Rock arrangements, mounds, plant combinations, groups of wild flowers and grassland may be used for decoration; these changing landscapes can leave motorway users with a lasting impression (Figures 4.48–4.50).

Figure 4.48   Natural type.

Figure 4.49   Natural type.

Figure 4.50   All kinds of herbaceous plants grow in the wide central reservation creating a very natural scene.

4.5   Landscape design and analysis of the central reservation of the Nanjing-Hangzhou motorway

In the light of the width of the central reservation and the ecological and biological characteristics of the anti-glare trees selected for the Nanjing-Hangzhou Motorway, eight greening methods have been designed for the central reservation.

Single row type of anti-glare *Sabina komarovii*

In the section which is about 10 km long from the Luojiabian Interchange to the toll station, the central reservation is 2 m wide. *Sabina komarovii* has been selected as the major anti-glare tree and planted in the middle of the central reservation at intervals of 2 m. *Ligustrum ovalifolium* Vicaryi and *Berberis thunbergii* have been interspersed between the curb and the median barrier. The ground cover plant selected is *Trifolium repens* (Figures 4.51 and 4.52).

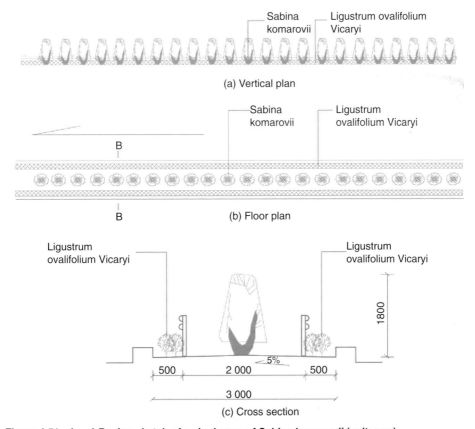

(a) Vertical plan

(b) Floor plan

(c) Cross section

Figure 4.51   (a–c) Design sketch of a single row of *Sabina komarovii* (unit: mm).

Figure 4.52 *Sabina komarovii* and *Ligustrum ovalifolium* Vicaryi.

It is appropriate to plant *Sabina komarovii* in a single row on narrow central reservations. It is also rational to make full use of the space by arranging coloured-leaf plants between the median barrier and the curb. *Sabina komarovii* stands vertical, with the bright yellow-red belt comprising *Ligustrum ovalifolium* Vicaryi and *Berberis thunbergii*. The ground cover plant selected is light green Trifolium repens. This results in a simple yet attractive green belt (Figures 4.53 and 4.54).

Figure 4.53 *Sabina komarovii, Berberis thunbergii* and *Trifolium repens.*

Figure 4.54 Single row of *Sabina komarovii.*

Interspersing *Osmanthus fragrans* and *Lagerstroemia indica*

Osmanthus fragrans is selected as the primary anti-glare tree, planted at intervals of 4 m. It is interspersed with *Lagerstroemia indica*. Four small *Osmanthus fragrans* are placed together to form a larger *Osmanthus fragrans*. *Ligustrum ovalifolium* Vicaryi and *Zephyranthes candida* are planted alternately between the median barrier and the curb (Figure 4.55) in a staggered manner. *Osmanthus fragrans* is a fine tree species used for landscaping in south Jiangsu Province, which is evergreen all the year around and blooms from mid-September to early October. Four-seasonal *Osmanthus fragrans* can blossom throughout the year. Its fragrant flowers provide a pleasant fragrance for travellers during their journey. *Lagerstroemia indica* is also a native plant that is adaptable to different environments and resilient to pruning.

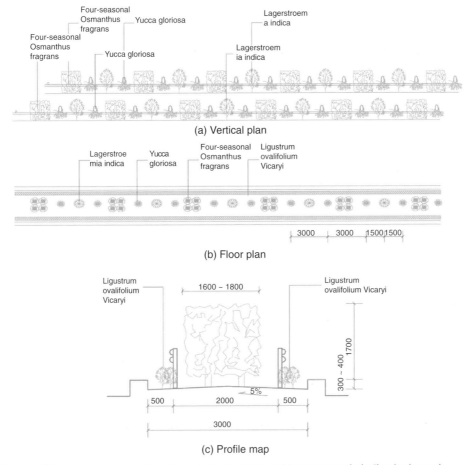

(a) Vertical plan

(b) Floor plan

(c) Profile map

Figure 4.55 **(a–c) Interspersing *Osmanthus fragrans* and *Lagerstroemia indica* (unit: mm).**

Its brightly coloured flowers bloom for a long period of time and add a significant amount of colour to the motorway in summer and autumn.

This design type is intended to introduce new types of primary anti-glare trees but in practice it has provenhard for *Osmanthus fragrans* to recover its abundant foliage after being transplanted. It is currently lacking in branches and leaves and not growing well. In some sections, its anti-glare effect is poor and the landscape is unsatisfactory.

Some *Osmanthus fragrans* were not placed together but rather planted independently. Those that were carefully planted and tended have grown better after being transplanted than those placed together as four plants but they still experience the problems mentioned above. Their appearance was later improved by adding *Canna indica* to both sides. The flowers of *Canna indica* are a fiery red, adding a significant amount of coloir to the green belt. In addition to *Lagerstroemia indica*, *Malus halliana* has also been selected to add a touch of spring scenery. Its beautiful and attractive flowers adorn the motorway in early spring.

Interspersing *Trachycarpus fortunei* and *Lagerstroemia indica*

Trachycarpus fortunei is chosen as the major anti-glare tree. Three trees of *Trachycarpus fortunei* are planted one after another before interspersing with one *Lagerstroemia indica*. *Ligustrum ovalifolium* Vicaryi and *Lorpetalum chinense* are arranged on both sides, and the ground cover plant is *Trifolium repens*. *Trachycarpus fortunei* is another precious greening tree species in the region south of the Yangtze River. Its trunk is tall and straight and its leaves are fan-shaped, beautiful and unique, and therefore it is a favoured greening tree species. This is particularly the case if it is combined with species such as *Yucca gloriosa* and *Musa basjoo* (Japanese banana), resulting in typical southern coconut island scenery. Choosing *Trachycarpus fortunei* as the primary anti-glare tree for the central reservation is a new innovation in central reservation design.

It is important to design carefully when selecting *Trachycarpus fortunei* as the major anti-glare tree. It is difficult for a single row of *Trachycarpus fortunei* to achieve the necessary anti-glare effect because it relies on its fan-shaped leaves to shield and block the light; a single row of *Trachycarpus fortunei* does not have a sufficient number of leaves for this purpose. According to the design construction effect (Figure 4.56), the anti-glare and aesthetic effects cannot be achieved, so this type of design requires further consultation. If two rows of *Trachycarpus fortunei* are planted, with one row higher than the other, or the two rows are combined with other tree species, then the above problem will be eliminated. More attention should be paid to the height of *Trachycarpus fortunei* in the central reservation. Its leaves should extend naturally by 1 – 1.5 m; if the leaves are too high or too low, they cannot shield glare. *Lorpetalum chinense* is planted between the *Trachycarpus fortunei*.

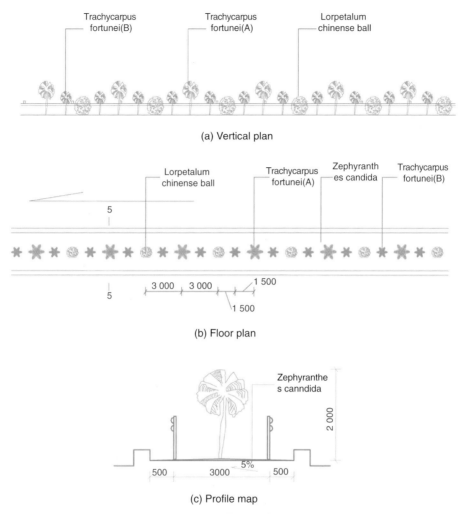

(a) Vertical plan

(b) Floor plan

(c) Profile map

Figure 4.56   (a–c) High and low _Trachycarpus fortunei_ (unit: mm).

In April, the flowers and new leaves of _Lorpetalum chinense_ are a very attractive purplish-red (Figure 4.57).

Interspersing _Juniperus chinensis_ and _Hibiscus syriacus_

In the central reservation of the Donglushan section, the initially selected primary anti-glare tree was _Nerium oleander_ but it was later replaced by _Juniperus chinensis_ due to its fluffy appearance and nonuniform size. The _Juniperus chinensis_ is combined with _Hibiscus syriacus_, which grows rapidly. The evergreen leaves of _Juniperus chinensis_ are much greener and more lifelike compared with _Sabina chinensis_. The beautiful flowers of _Hibiscus syriacus_ appear in summer. _Ophiopogon japonica_ acts as an attractive and green ground

Figure 4.57 *Trachycarpus fortunei* **interspersed with** *Lorpetalum chinense.*

Figure 4.58 **Interspersing of** *Juniperus chinensis* **and** *Hibiscus syriacus.*

cover plant. Golden yellow *Ligustrum ovalifolium* Vicaryi is planted beside the curb, adding fresh and distinctive colours to the whole green belt (Figure 4.58).

Double row of *Sabina komarovii*

The primary anti-glare tree, *Sabina komarovii*, is distributed in two alternating rows within the median barrier. For single rows, the trees are planted at intervals of 3 m; for double rows, one *Lagerstroemia indica* is planted after three *Sabina komarovii* and three *Yucca gloriosa* have been planted at equal intervals between two *Lagerstroemia indica*. *Lorpetalum chinense* or *Ligustrum ovalifolium* Vicaryi is densely planted between the median barrier and curb. *Trifolium repens* is the ground cover plant. The end of the safety island is 8 m in length, which is densely planted with *Sabina komarovii*. Their tops are pruned into a sloping formation. A section of this type is 10 km in length (Figure 4.59).

(a)

(b)

Figure 4.59   (a, b) Double row of *Sabina komarovii*.

Square column-shaped *Viburnum odoratissimum*

Viburnum odoratissimum (1.6–1.7 m in height and 1 m × 1 m in crown diameter) is planted at intervals of 1.5 m on the middle line of the central reservation. Between two trees, three *Yucca gloriosa* are planted; *Trifolium repens* is the ground cover plant. *Lorpetalum chinense* of 31–40 cm in height is densely planted between the median barrier and the curb. The end of the safety island is 8 m long and densely planted with *Viburnum odoratissimum*. Their tops are pruned into a sloping formation. A section of this type is 10 km in length (Figure 4.60).

(a)

(b)

Figure 4.60   (a, b) Square column-shaped *Viburnum odoratissimum*.

Sabina komarovii on one side of the strip

Sabina komarovii is selected as the primary anti-glare tree species and planted on one side of the median barrier. It is planted at intervals of 1.6 m; the unit length of this type is 10 km. On the other side, three *Hibiscus syriacus* are planted side by side at intervals of 70 cm. The interval between two *Hibiscus syriacus* groups is 1.8 m. *Ligustrum ovalifolium* Vicaryi is densely planted between the median barrier and the curb; *Sabina komarovii* is arranged on the other side. On the opposite side, *Pittosporum tobira* is planted at intervals of 3.2 m and interspersed with *Yucca gloriosa*. *Lorpetalum chinense* is densely planted between the median barrier and the curb. The plants are therefore interspersed. For the 8-m-long section of the safety island end, *Sabina komarovii* is densely planted with the tops pruned into a sloping formation. The ground cover plant is *Trifolium repens* (Figure 4.61).

Figure 4.61   (a–c) *Sabina komarovii* on one side of the strip.

Shutter blind type *Viburnum odoratissimum*

Viburnum odoratissimum is chosen as the primary anti-glare tree, which is planted in a row at intervals of 3 m and at an angle of 45° to the motorway. It features a sloped design and the trees are 1.6 – 1.8 m in height. *Buxus sinica* is planted densely between the median barrier and the curb, and *Trifolium repens* is the ground cover plant. The end of the safety island is 15 m long and densely planted with *Viburnum odoratissimum*. Their tops are pruned into a slope and the periphery is densely planted with *Buxus sinica* (Figure 4.62).

Figure 4.62   Shutter blind style *Viburnum odoratissimum*.

5 Environmental Landscape Design of Service Areas and Toll Stations

5.1 General

Motorways are a product of economic development and also represent a country's level of modernization. Cities have been connected by transport infrastructure ever since they came into existence. In the 21st century, motorways will become the main way to connect cities together. In the wake of cities' economic development and the modernization of transport infrastructure, motorways have developed rapidly in China. However, motorways have changed and even damaged the environmental characteristics of regions along motorways, affecting the balance of nature and resulting in increasing conflicts with the environment. In addition, people now demand more cultural and artistic features along motorways following an increase in their level of education. How to go about integrating constantly developing road transport with the surrounding environment and build a safe, fast, comfortable and attractive route have become crucial issues in motorway construction today. The overall design concept of buildings and landscapes along the Nanjing-Hangzhou Motorway is the so-called 'pearl necklace' concept. A brief outline will be given of the development of this concept and providing an analysis and summary, as well as discussing how the Nanjing-Hangzhou Motorway managed to achieve the general design requirements of 'scenery, tourism, nature and environmental protection'.

Factors contributing to the concept - spatial environment formed by the motorway

Roads can be likened to a city's corridors and windows, acting as the main way for people to see and understand a city. The road environment is an important component of the urban environment and also accounts for one of the city's three types of spaces (traffic, building and open spaces). This ribbon-shaped

The Environment and Landscape in Motorway Design, First Edition.
Qian Guochao, Tang Shuyu, Zhao Min and Jing Chun.
© 2014 China Communications Press. Published 2014 by John Wiley & Sons, Ltd.

environment is an important reflection of a city's appearance and character. In *The Image of The City*, Kevin Lynch (Lynch, 1960) argues that roads are the most important of the five factors constituting a city's image, indicating the importance of road landscapes in creating a characteristic city image.

Motorways are entirely enclosed major transport routes that link cities together. Their aim is to increase the speed and flow of traffic; vehicle speed is generally in excess of 80 km/h. Motorways are also open spaces with linear, directional, continuous and fast-moving features. Due to their function and geographical position, motorways form a spatial environment with different visual, environmental and geographical features to urban roads.

In light of this concept, building facilities located between cities along the motorway play a very important role in the maintenance and development of a city's image as well as the entire region's economic and cultural development.

Visual features

Visual landscapes are based on people's perception of a visual image, using physical scenery to create a visually appealing environment according to the laws of aesthetics. The visual features within the spatial environment of the motorway must be analysed in order to find a pattern, which is then used as one of the bases for landscape design.

We all know that the vehicles travel at a high speed on the motorway and drivers and passengers in vehicles can be seen as viewers of landscapes along the route, and when vehicles travel at high speeds, the head of a driver in a vehicle moves only in a very small range and his/her sight focuses on the lane ahead. The driver tends to focus on the same point and his/her field of vision is narrow, resulting in so-called tunnel vision. However, the passengers' field of vision is much greater, meaning that they can view the surrounding landscape from many different angles.

Data shows that motorways have the following visual features:

(1) Vision. Vision weakens as vehicle speed increases. At a moderate speed it takes drivers and passengers 1/16th of a second to see a target clearly. In between these targets, vision is blurred. Drivers therefore only have one chance to view a particular scene clearly.

(2) Visual field. As vehicle speed increases the visual field becomes smaller. The distance between points of visual fixation increases and the distance needed to identify the point ahead decreases. For example, at a speed of 70 km/h, the point of visual fixation is 360 m ahead of the vehicle and the field of view is 65°; at a speed of 100 km/h, the point of visual fixation is 600 m ahead of the vehicle and the field of view is 40°.

(3) Adaptability to brightness. People initially see nothing when entering a dark place from a bright environment. After their sight has adapted to the dark environment, they can distinguish the outline of an object. This process of adaptation is called dark adaptation. Entering into a bright

environment from a dark place, can be very blinding at first and surrounding objects cannot be seen clearly. After adaptation, vision then recovers to a normal state. This phenomenon is called light adaptation. When entering and exiting tunnels the level of brightness can change abruptly, and people's vision cannot adapt so quickly to the change, meaning that drivers cannot see ahead clearly. This is known as the so-called 'black hole effect'.

(4)   Colour sense. Red is of highest visibility, followed successively by orange, white and green.

(5)   Glare. Very bright objects or a strong brightness contrast within the field of vision can be uncomfortable for the driver or result in deterioration in vision. The headlights of vehicles driving in opposite directions create glare, which often means that surrounding objects cannot be seen clearly. In severe circumstances, glare can be painful or result in deteriorated eyesight.

(6)   Visual fatigue. Monotonous scenery frequently causes visual fatigue when driving for long periods.

Natural environment

The natural environment includes factors such as topography, climate, vegetation and surrounding environment. Although motorway construction has greatly promoted national economic development, it has also resulted in serious environmental problems. Motorway structures damage the appearance of the surroundings and increase radiant heat. Furthermore, original natural vegetation and soil on slopes are destroyed during construction, which results in a hostile environment as well as infertile, dry and unstable soil, posing many challenges for slope greening. These include construction difficulties, low vegetation survival rates and tending and management challenges.

The climate, hydrology and site conditions in different regions along the route directly affect plant growth. When the motorway is in operation, noise, vibration and exhaust emissions from fast-moving vehicles as well as heat absorption and radiation from the road surface in summer are detrimental to the natural environment and not conducive to plant growth.

Regional culture

Culture has a major role to play in ensuring the cohesion of a nation. As a vast country with a long history, China has many historical records as well as a rich cultural history. Relics of China's 5000-year-old civilization can be found all over the country.

The formation of a regional culture is influenced by factors such as the surrounding geology, topography and climate. Long-term social development eventually results in different regional cultures. Different regions have their owncultures; each culture has its own individual character and identity. The Jiangsu section of the Nanjing-Hangzhou Motorway is nearly 200 km long, passing through different regions. It is therefore a good tool for exhibiting and promoting different regional cultures. Depicting these regional cultures will also enrich the cultural aspect of the motorway landscape.

Regional culture relates to natural scenery, national customs, religious beliefs, cultural relics, historical sites, folk art and historical figures. Natural scenery is the primary constituent of motorway landscapes but beautiful scenery alone is not sufficient and culture also has a major role to play to prevent the landscape from becoming monotonous. During the landscape design of the Nanjing-Hangzhou Motorway, culture is used as a guide to research and create a landscape environment rich in regional culture. This can help drivers and passengers better understand the local history.

Composition of the concept – landscape elements

(1)   Service areas. Service areas directly reflect the level of a motorway's services. They also represent comprehensive user-friendly design.'(2) Branch management centres. These can be likened to a motorway's 'central nervous system'. They are an important component of motorway management.

(2)   Toll stations. This refers to internal and external environments, including toll gates and management buildings.

(3)   Tunnels. The design of tunnels involves a combination of their practical function and appearance, using artistic means to improve the shape and colour of tunnel entrances, as well as using decoration and lighting to solve the 'black hole effect and make for a more pleasant driving experience.

(4)   Greening. Greening is the most intuitive and widely used element in the building area. It directly affects the overall landscape and plays a critical role in improving the environment.

(5)   Subsidiary facilities. These include curbs, peripheral walls, noise barriers, median barriers, road signs, furniture, flower beds and small sculptures, which can fully reveal the beauty of the landscape in the building area.

Development of the 'Pearl Necklace' concept – dynamic sequential layout of landscapes

People travel at a high speed when on motorways, meaning that landscapes pass by in an instant. However, the landscapes are continuous, rather like a moving sequential exhibition of visual pictures. There should be change but also uniformity in the landscape as well as flat and undulating areas that reach a climax. The building facilities on the motorway constitute this climax.

Drivers and passengers will pass by many sections during their journey, and this of course includes many motorway buildings between cities. These buildings will naturally leave people with an impression. If these impressions have no continuity and there is no connection between them, then the facilities cannot become an entity and the landscape will be incomplete. There must be a beginning and an end with a climax in between. The landscape must feature change and not be monotonous. A continuous landscape without a pattern cannot achieve a diverse and uniform artistic effect. The overall landscape design of the Nanjing-Hangzhou Motorway has focused on the 'pearl necklace' concept, whereby the motorway is likened to a shining pearl necklace. Important structures along the motorway, such as interchanges, and important buildings, such as service areas and management centers, form the pearls on the necklace. Each sparkling pearl then is strung together to form unique patterned and continuous landscapes.

The overall design of the building areas on the Nanjing-Hangzhou Motorway therefore ensures uniformity to improve their appearance. Each individual unit varies slightly whilst guaranteeing overall uniformity. This means that each individual unit can combine the environmental features of each motorway section and develop successfully.

Design principle of the 'Pearl Necklace' concept

The landscape of the Nanjing-Hangzhou Motorway is designed to stimulate people's visual attention. During landscape design, different elements should be combined according to the environment as for the Nanjing-Hangzhou Motorway. The aim should be to create a road landscape that features 'scenery, tourism, nature and environmental protection'. In order to achieve this, the following principles should be followed:

(1) Guarantee the efficiency and traffic safety of the motorway.
(2) Fully consider the visual characteristics resulting from driving at high speed and avoid becoming caught up in detail due to focusing too much on technique and appearance.
(3) Respect nature and comply with the requirements for environmental protection and nature. Make up for and repair the damage caused during motorway construction by combining eco-friendly construction with environmental protection measures.
(4) Combine and exhibit regional cultures along the route and respect local customs and traditions.
(5) Use art and innovative thinking in order to create an attractive and charming 360° landscape and develop the local scenery.

5.2 Building works and the 'Pearl Necklace' concept of the Nanjing-Hangzhou motorway

On the Nanjing-Hangzhou Motorway, the 'pearl necklace' concept is used mainly for important structures along the motorway, such as interchanges, flyovers, tunnels, service areas, management centres, toll stations and toll gates. Only when each individual 'pearl' is exquisite can the whole 'necklace' shine. In this section the five most impressive building works are discussed in order to describe in more detail the 'pearl necklace' concept. The five considered are: the toll station on the main route; the Donglushan Service Area; the Management Centre of the Nanjing-Hangzhou Motorway; the Tianmuhu Lake Service Area; and the Taihu Lake Service Area.

Toll station on the main route of the Nanjing-Hangzhou motorway

The whole toll station and its roof look like a silver wing flying over the edge of the motorway. This toll station is the starting point of the Nanjing-Hangzhou Motorway and therefore also the first 'pearl' on the 'necklace'. Its roof rises up from one side of the motorway, imitating the silver airplanes taking off and landing at the Nanjing Lukou International Airport nearby. The wing-shaped roof provides a sculpture-like canopy for the toll station and is an important and unforgettable landmark and gateway to the airport region. The toll station fully reflects the imposing and high-speed characteristics of the motorway (Figure 5.1).

The toll booths of the toll station are located below the central axis of the management building unit, which is made of a steel framework covered with pure aluminium plates painted with fluorocarbons, mimicking aeroplane construction technology. The broadest space in front of the management building is the monitoring room, which is higher than the management office and meeting facilities, both of which are arranged at either side of the toll plaza for monitoring and guiding the work of the toll station, as well providing a buffer and transition between the noisy motorway and the staff

Figure 5.1   Overview of the toll roof seen from the motorway.

dormitory and other living facilities. The monitoring room and neighbouring office are covered with soundproof glass and the side wall of the meeting room is an extension of the roof of the whole management building; the roof curves downwards until it reaches the ground, forming a silver aluminium wall (Figure 5.2).

Using the stairs within the semi-outdoor space of the management building leads to the entrance of the staff living quarters opposite; looking further back, there is an area of greenery behind the station area. Seen both close-up and from a distance, the beautiful landscape makes the internal and external spaces as well the buildings blend perfectly into the surrounding environment. The staff dormitory is a single-storey building, with the staff canteen on one side. The canteen wall is an extension of the roof of the whole dormitory building, which curves downwards until it reaches the ground, just like the meeting room of the front management building. Because the shape of the roof of the management building and living quarters is identical, they appear to be one

Figure 5.2   **Monitoring room of the management building.**

Figure 5.3   Horizontal view of the whole 'wing' seen from the canteen in the living quarters.

single unit when seen from a distance on the motorway, jointly forming a 'silver wing' with the roof of the toll station at the front (Figure 5.3).

When designing the whole construction drawing, the most challenging and difficult part was to design the steel structure of the toll station roof. The toll station roof on the main route is very difficult to design because it is a hyperboloid structure and has a large cantilever and thin components. There are four difficult points: first, selecting an appropriate calculation mode for the structure; secondly, there is a large span and few supporting points as well as strict restrictions over the height of components and pillar sections due to the shape; thirdly, treatment of the support joint; and fourthly, treatment of the juncture between the roof and the office building.

There are six incoming and eight outgoing toll lanes at the toll station, and the area of the toll station is $1780\,m^2$. The toll office building covers an area of $1030\,m^2$, and other auxiliary buildings cover an area of $700\,m^2$. In order to meet architectural requirements and facilitate the movement of the staff, the toll station is connected to the office building. The building complex is $102.8\,m$ in total length, $61.5\,m$ wide at its widest point and $11.2\,m$ above the road surface at its highest point.

The toll roof has six supporting bases in total, each of which is supported by two concrete filled 351# steel tubular columns. The distance between the columns is $21.9\,m$, and the distance between one end of the toll canopy and the office building is $30.2\,m$; at the other end are cantilevers, of which the longest is $17.20\,m$ (Figure 5.4).

Two sheets of semi-transparent fibreglass, the first of rectangular shape with dimensions $46.5\,m \times 3.8\,m$ and the second of elliptical shape with dimensions $17.8\,m \times 3.8\,m$ are placed on the central axis and sides of the roof to make the toll station appear more innovative. In order to protect passing vehicles from

Figure 5.4   The longest cantilever of 17.20 m.

the impact of falling rainwater that has accumulated on the roof, drainage ditches have been constructed on both sides of the roof, meaning that some of the water on the roof is channelled away and the rest drains away naturally.

The main body of the office building is made of deformed reinforced concrete frame columns with cast-in-place slabs. On the roof plate, square steel stringers are mounted on the welded H-shaped girders, on which aluminium sheets are laid in the same way as the roof. This is shown in Figure 5.5.

The building plane of this work is a hyperboloid space steel structure; there are currently no similar examples or technical data available in China, so this

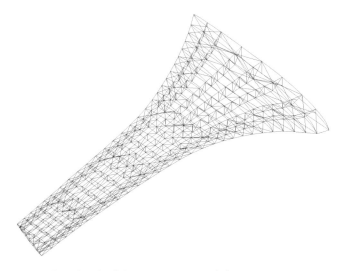

Figure 5.5   Design sketch of the canopy space girder.

resulted in many difficulties during the design of the structure. Examples of these difficulties included what structural form to use, how to arrange the structure, which calculation method to use, treatment of the juncture between the steelwork roof and the concrete structured office building and which supporting structures to use. Many comparisons were made between different methods during design.

Selection of structure system

The roofs of motorway toll stations are generally made of steel truss structures but this was not suitable for this work due to dimensional restrictions, in particular due to the long and thin cantilever which is only 20–60 cm thick. After repeated comparison, a space steel tube girder structure was finally chosen. The longitudinal girder is the main girder and the horizontal girder (along the axis of the motorway) is the auxiliary girder. The support and purlin function to enhance the overall stability of the structure. The girder is made of steel tubes. Continuous connection is ensured between steel tubes to reduce the sectional dimension of the structure and meet architectural requirements.

Arrangement of the structure

The plane and section of the roof are irregularly cambered. The central axis of the canopy is first aligned with the center of the office by 1–6 axes. Then both sides are extended from here before arranging 5 vertical primary girders and 2 vertical secondary girders. Then 24 horizontal secondary girders are arranged on the longitudinal girders. Then necessary diagonal bracings are arranged on the upper and lower cords of the girders; all the primary and secondary girders are connected at an equal height to form the body of the space girder structure. The root of the primary girder adjacent to the office building is 2–4 m in height and varies with the curve; it is 0.2–0.6 m in height at the end of the cantilever (Figure 5.6).

Because the upper and lower envelopes of the whole roof are made of aluminium sheets, there is a stressed-skin effect. The secondary member is strong enough to restrict the lateral displacement of the primary member, so the supporting arrangement of this structure is relatively simple.

Treatment of junction

At the beginning, one end of the roof girder is supported on the frame column of the office building (1-B); because the horizontal thrust of the girder is too large, a frame column and beam cannot be built within the architectural dimensions allowed, meaning that the main bodies of the roof and the office building have to be separated to form their own system. A row of square concrete filled with steel tubular columns is added at the junction to guarantee the foundation position of the roof column. The office building is treated with a structural transition layer, using beams to support its columns. This solves the problem of superposing of the two foundations. The roof plates are made of

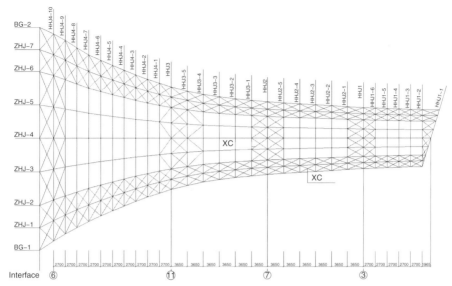

Figure 5.6 **Arrangement plan of primary and secondary girders (unit: mm).**

aluminium sheet and their deformability is strong, so they may be connected into a whole without considering jointing.

Selection of materials

The primary girder is made of hot-rolling seamless pipe Q345B, and the secondary girder, strut, purlin and steel tube are made of Q235B. Q235B is welded with the manual welding rod E4303, and Q345B with E5016. H-shaped steel is connected by high-strength bolts of grade 10.9 and others by plain bolts of grade C.

Construction measures

(1) Considering the roof has a large cantilever and the structural section is not very high, for example ZHL-7, the girder at the supporting base is only 612 mm high and the cantilever is as long as 16 526 mm. In order to reduce the deflection, the long cantilever girder is filled with steel plates. Meanwhile, arching is necessary to prevent the 'aeroplane wing' from falling down.

(2) Two nodes near the girder of the supporting base are also filled with steel plates to enhance the rigidity.

The unique bi-directional, double-faced, bending, large span cantilever steel structure on the main route of the Nanjing-Hangzhou Motorway has already been completed and put into use. The 'silver wing' now 'flies' above the imposing motorway appearing to keep an eye on the traffic flow and symbolizing the rapid development of motorways in Jiangsu Province (Figure 5.7).

Figure 5.7 'Silver wing' shining in the sunlight.

Donglushan service area

The Donglushan Service Area is located in Lishui County, Jiangsu Province, and the scenery of the surrounding region is beautiful. The area features undulating hills, crisscrossing water networks, large and small tea plantations and bamboo groves in the mountains and plains.

Even during the design and preliminary design stages of the project, high requirements were set for the building and landscape design of this service area. According to the requirements, this service area did not only need to meet the service demands of passing vehicles, drivers and passengers but it also needed to reflect the overall design requirements of the Nanjing-Hangzhou Motorway and its 'pearl necklace' concept of 'scenery, tourism and nature'.

During the project planning, project design, preliminary design and construction drawing design, the architects deeply studied and comprehensively implemented the overall design requirements and concepts, beginning with a user-friendly design concept. They expressed the design ideas of 'scenery, tourism and nature' in various ways in order to reflect the 'pearl necklace' concept. A user-friendly design means implementing a design that 'puts people first' during the whole design process as well as fully respecting and paying attention to the needs of users.

User-friendly selection of location

First, architects actively work with the construction unit and motorway design department during the location selection process to carry out an in-depth study on natural conditions such as the landscape and topography along the route, making comparisons between different candidate sites.

According to the conventional method of site selection for the motorway service area, a service area should usually be located in a flat area along the motorway route and service facilities should be arranged on both sides of the motorway at the same location. Based on the construction requirements for service facilities, the distance between two service areas should generally be about 50 km. It is therefore necessary to build a new service area in an appropriate area near to Section K50-K55 on the Nanjing-Hangzhou Motorway. However, this section is a hilly area with an undulating topography, so no plot has been found that is big enough to accommodate the service area. For this reason, the architects, in consultation with the motorway design authority and construction unit, changed the previous site selection method for service areas. They have developed a brand new user-friendly model for selecting the site of the service station, guaranteeing that its functions will not be affected.

Based on detailed research of a topographic map along the route and on-the-spot surveys, the architects, construction unit and motorway design authority worked together and determined that the service area would be located between Section K53+200-K53+700 on both sides of the Nanjing-Hangzhou Motorway. On the north side of this section, it is unsuitable to build large-area service facilities due to undulating topography; on the south side, however, the natural conditions are excellent; on the one hand, the terrain is flat and there is a large plot of land; on the other hand, there is a large natural reservoir about 350 m away. The reservoir is surrounded by undulating hills, verdant trees and picturesque scenery, meaning that this is an appropriate location to establish the service area. According to the user-friendly method for selecting the site of service stations, this area is very suitable as long as the relationship between the hilly north and flat south areas is handled well in design. In the future design, the major service facilities may be concentrated on the south side, whereas there may be only a few essential facilities on the north side. The north and south areas will also be connected to each other. This type of site selection meets the requirements for the distance between motorway service stations on the one hand and also provides a favourable environment for future buildings and landscape design. Architects can make the future building design and landscape more attractive by making full use of the surroundings and the user-friendly design concept (Figure 5.8).

User-friendly general layout

The north service area covers a small area due to the limitations of topographical conditions (the site and its surroundings are very hilly and the difference between the highest and lowest points is 16 m), so an area with a small elevation difference has been selected as the site to build a service area with a gas station, toilet and limited parking space. It covers an area of 40 mu (about 26 667 m^2). The south service area has favourable natural conditions with the elevation difference being small within a large area and adjacent to a

Figure 5.8 Design of the Donglushan Service Area.

large reservoir. This means if the main service facilities are built here then the distant mountains and near water may contribute to the scenery. In the design plan, the south service area covers a larger area of 85 mu (about 56 667 m^2). A large parking area, dining area, rest area, large toilet area and petrol station will be built here as well as subsidiary facilities such as management building, pump room and distribution room. Because this section is in an excavated section, the natural topography on both sides of the motorway is higher than the motorway. The north and south service areas are therefore connected by an attractive flyover (Figure 5.9). Such a layout overcomes the adverse impact of a big elevation difference on the layout of the service area and reduces large-scale earthwork, avoiding damage to the original natural topography as far as possible. On the other hand, this kind of layout takes full advantage of

Figure 5.9 Attractive flyover in the service area.

the original natural landscape resources along the motorway, allowing motorway users to enjoy the surrounding natural scenery, which fully reflects the philosophy of user-friendly design. The north service area has fewer service facilities and there is a gas station with a toilet near the exit. Passing vehicles, drivers and passengers can easily get to the south area to access the service facilities there via the attractive flyover.

The main service facilities of the service area are concentrated in the south area. The general design focuses on a user-friendly concept.

First of all, the main building (comprehensive service building) is arranged towards the south of the middle of the area. This place is the highest natural point in the south service area, meaning that the main building occupies a central position; on the other hand, it also ensures that the main service facilities are closer to natural water. Passing visitors can therefore admire the distant landscape when using the service facilities. The building is in a curved shape; the main entrance and the toilet entrance face northwards towards the motorway in the distance and are located next to a parking area. The south side of the main building faces the natural reservoir and distant mountains, where a rest platform has been built along with a bridge over the water (Figure 5.10).

The parking area in the south service area includes parking spaces for minibuses, coaches and lorries. One side of the minibus and coach parking areas is close to the entrance ramp, and the other side is next to the main building, while the lorry parking area is adjacent to the exit ramp. The different parking areas are clearly divided by a large amount of grass, trees and shrubs. Such a layout means that it is convenient for visitors to use the service facilities and avoids visual interference, creating an excellent parking environment and reflecting the user-friendly design concept.

Figure 5.10 Outdoor platform and bridge.

There used to be a small pinewood in the northeast of the south area with beautifully shaped pines. For the sake of nature and environmental protection, the architects conserved the small forest as far as possible during design, making for a more natural landscape in the parking area.

The subsidiary rooms (power distribution room, pump room and vehicle maintenance room) are arranged to the east of the main building, hidden away by grass, trees and shrubs. The petrol station is designed at the exit of the area for the convenient refuelling of vehicles.

User-friendly design of the main building

The main building (comprehensive service building) is curved (Figure 5.11). The rest area and dining area are arranged to its south and make for a pleasant view with the outdoor natural reservoir, making use of and borrowing from the surrounding scenery. The main entrance and toilet entrance are arranged on the north side, facing the motorway and parking area (Figure 5.12). By applying the user-friendly design concept, a large artificial stone wall has been erected facing the motorway and next to the parking area and separates the rest area from the busy motorway. The side of the building facing the natural reservoir is made of large sheets of transparent glass, enabling visitors to face the natural water and distant mountains when taking a break or having a meal. This design method therefore fully respects and considers the needs of users (Figure 5.13).

Different functional areas of the building have been appropriately divided. Busier and quieter areas have been separated naturally and the flow of passengers, staff and goods is clear and straightforward. As a modern service building, the user-friendly design concept has been fully considered.

Figure 5.11   Distant view of the main building.

Figure 5.12   Main entrance of the comprehensive service building.

Figure 5.13   Restaurant and outdoor waterscape.

The main building is streamlined with the main entrance area raised; the main entrance and the area around the large toilet area have been paved with L-shaped sheets. The main entrance area is clearly visible due to the mature, imposing and simple design.

The interior and exterior decorative materials of the building are natural and simple and the colours complement each other well. The design reflects the simplicity of modern service facilities and makes the buildings integrate into the surrounding natural environment.

East Lishui Branch Management Centre

The East Lishui Branch Management Centre is an important architectural building project on the Nanjing-Hangzhou Motorway. The designers have focused on the requirements of the 'pearl necklace' concept, ensuring that the building blends into the surroundings.

Looking at the East Lishui Interchange from a distance along the motorway, a group of simple and unadorned buildings with blue tiles and white walls can be seen amongst the trees, blending seamlessly into the surrounding environment. Only when visitors enter the area can the buildings' charm be felt. There are a plethora of different architectural styles in addition to green trees, grass and rivers, making the visitor feel as though they are in a large garden. Visitors are left feeling relaxed and content on account of the fusion of the buildings and the surrounding environment (Figure 5.14).

Relationship between buildings and environment

The Lishui Branch Management Centre is located at Section K1+149 on the Nanjing-Hangzhou Motorway in Lishui, where the terrain is slightly hilly. The whole interchange ramp area has been built in an area of expropriated land. The surrounding natural environment is beautiful, with an elevation of 30–36 m and no major obstructions. In such an environment, it is important for designers to consider how to integrate the buildings into the natural surroundings and bring the environment to life.

The Lishui Branch Management Centre is composed of the management office building, toll office building, living quarters and subsidiary buildings. It covers a total area of 5217 m^2 (Figure 5.15). The management office building is the most important and also the largest building in the complex; it has been built as far back as possible so that the area at the front can be used as a square

Figure 5.14   Lishui Branch Management Centre seen from the Nanjing-Hangzhou Motorway.

Figure 5.15   Main building of the East Lishui Branch Management Centre.

within the management centre. The front square is separated from the toll station by a green landscape. The spacious internal square is both the entrance to the main building and a parking area for cars needing to access the management office building. It also provides sufficient distance in front of the main building such that the entire main building can be seen. The toll office building is arranged at the front of the area next to the exit of the toll square so that the toll station is able to monitor the surrounding area effectively. There is a small parking area in front of the office building for staff. The living quarters are located well away from the motorway to the left and back of the management building. Although it is separate from the management building, it is still very convenient to reach. The Lishui Branch Management Centre's most distinct feature is a secluded river that passes by the three individual buildings, each of which faces towards the river. For this reason, the three seemingly separate buildings form an integrated unit with the river and road (Figure 5.16).

The buildings are concentrated together to form an architectural complex. The surrounding environment adds contrast to the complex making it stand out. When travellers pass by the east Lishui Interchange ramp, they can admire a charming blue sky landscape surrounded by green hills and clear water.

Figure 5.16   Buildings are connected by the river.

Relationship between buildings

The three individual buildings in the East Lishui Branch Management Centre have their own characteristics whilst maintaining a common style. The bluish-grey roofs, light-coloured wall surfaces and grey plinths are the defining features of the buildings. The floor area of the management office building is 2560 m² spread over three storeys; its total length is approximately 75 m and the foundations are roughly rectangular. The entrance platform is 9 m wide and guides the flow of people into the main hall. Inside there are three staircases that add to the imposing style of the main building. The 5.4-m-high monitoring centre on the first floor and the activity room on the third floor have spacious dimensions of 18 m × 15.3 m and 15.6 m × 13.8 m, respectively. Their roof trusses have been made with cast-in-place concrete technology to good effect. The exterior styling of the building is simple, consistent and imposing and the regular shape of this building also gives people a sense of direct and explicit mass. The thick and solid awning at the middle entrance forms a strong visual centre and also adds to grandeur of the building. The stand column at the entrance is conical, which draws on a technique of Tang Dynasty ancient architecture, resulting in an austere effect. The whole building has a character that is 'simple, austere and grand', reflecting the characteristics of the management office building (Figure 5.17).

Compared with the management office building, the toll office building has a rather more lively character. Although they are similar in appearance, they have different purposes (Figure 5.18). The toll office building has a floor area of 1472 m² spread over three storeys and with an irregular floor design. There is a meeting room and activity room at both ends of the first floor and the remaining rooms are offices. In order to guarantee that the monitor room is built at sufficient height, the second floor is designed slightly higher and features attractive offices. The triangular roof trusses form a continuous pattern and lighten the atmosphere of the building. The irregular floor design, stepped

Figure 5.17 Entrance to the main building.

Figure 5.18   Toll office building.

roof, hollow roof trusses and exquisite handrails reflect the building's natural style.

The smaller living quarter has a floor area of 1025 m^2, wherein there is a canteen, restaurant and dormitory. As it is a service building, it has been designed as a spacious L-shape to make the building appear friendlier (Figure 5.19).

The three individual buildings form a uniform series: in terms of style, the buildings feature traditional sloping roofs whilst employing triangular roof framework and innovative details to create a simple and natural group of buildings. Orderly and tall edges and corners as well as simple and clear lines increase the modern look. The buildings feature their own distinct 'simple, exquisite and intimate' characteristics. Different architectural styles mean that the attractive buildings are full of diversity and blend very well in to the beautiful natural scenery. The buildings have a unique appeal and form a special relationship with visitors.

Relationship between buildings and culture

Built in the Sui Dynasty, Lishui, also known as Zhongshan in ancient times, has a long history. Du Mu, a poet in the Tang Dynasty, extolled the Qinhuai

Figure 5.19   Living quarters.

River with the words 'The evening mist is enshrouding the river, and the moon is casting its light on the sand, so I shall moor my boat and lodge at an inn for the night'. The wide Qinhuai River is like a Jiangnan landscape painting being rolled out slowly; it is also home to the fascinating and charming Nanjing Qinhuai culture. Lishui is a 'bright pearl' that has been inlaid at the source of the Qinhuai River. With blue tiles and white walls, a sloping roof and a wood-colour outer wall, the East Lishui Branch Management Centre is a reflection of local characteristics. Graceful areas of water and delicate little bridges improve the appearance of the landscape. The building blends very well into the environment, reflecting not only the local history and culture but also a very modern architectural style.

People need to communicate with each other; likewise, buildings also need to communicate with the environment. People feel a sense of happiness and identity when there is harmony between buildings and the environment. This is the kind of effect that architects strove to achieve when designing the East Lishui Management Centre.

Tianmu Lake Service Area

General

The Tianmu Lake Service Area is part of the Nanjing-Hangzhou Motorway project and plays a role in the middle of the 'pearl necklace'. It has many features. In light of the overall design requirements, it was originally designed as a parking area. According to the actual needs, some service functions will be added to the original parking area to form a small-scale service area. Its land and floor area are small. The Tianmu Lake Service Area is just 10 km away from the national AAAA Tianmu Lake Scenic Spot. The construction unit and the local authority hope to see its completion to be deemed as a successfulexample of architecture.

Architects must therefore plan the site accordingly and carefully consider the functional layout, mass, shape, space, material and colour of building according to the overall design requirements, ensuring that this small-scale public building possesses its own charm (Figure 5.20).

Overall layout

The Tianmu Lake Service Area is located symmetrically on both sides of the Nanjing-Hangzhou Motorway in Liyang (the Jiangsu section) at Section K97+700-K98+200, with an underground passage connecting the north and south areas. The planes of the north and south areas are trapezium-shaped (the upper base is 160 m, the lower base 330 m and the height is 120 m). The area, previously a farmland, is flat without any obstructing features. Its original elevation was about 4.6 m; the designed elevation after backfilling is 5.1–6.0 m.

Figure 5.20   Design of the Tianmu Lake Service Area.

The vehicles from the main route enter into the service area via ramps at both ends. Considering the high speed of vehicles on the main route, the service areas have been built 20 m further back and a green isolation strip and a through lane have been designed. The rectangular parking area is positioned close to the through lane for convenient parking and passing. Both ends of the parking area are next to the entrance and exit in order to make entering and leaving the parking area as convenient as possible. In order to facilitate the use of service facilities (taking the north service area as an example), the comprehensive service building is falcate-shaped (i.e. rectangle + semicircle) and located behind and directly next to the parking area. It lies in the centre of the area such that the comprehensive service building faces the main entrance. In addition, the semicircular part of the main building naturally encircles a semicircular outdoor square, and its main entrance is opposite the parking area. In order to enrich the landscape, a fountain is located at the centre of the semicircular square.

For convenient use, the petrol station is arranged close to the exit of the service area and the maintenance room is designed to its northwest. Subsidiary facilities, such as water, power and wastewater treatment facilities are located further back in the service area. The south service area is approximately symmetrical with the north service area.

Main building design

With regard to an individual service area, the 'pearl necklace' concept for the design of the Nanjing-Hangzhou Motorway emphasizes that the most important landscape is the main building of the service area (comprehensive service building). The design of the main building is therefore of great significance.

From the point of view of architects, modern architecture must be function-based, especially modern public buildings with specific functions. Once these functions have been assured, full consideration should be given to the mutual relationship between the mass, shape, space and colour of the building and the surrounding environment, as well as other influences from the regional climate, culture and traditions. An appropriate architectural style should be selected that blends into the surrounding environment and enhances the appearance of the landscape. This demonstrates the respect that the architects have for the surrounding environment. The main purpose of a motorway service area is to provide a comfortable resting place for travellers as well as convenient dining and toilet facilities and fast refuelling and vehicle maintenance services. The building and landscape design of the service area should create a relaxed and comfortable atmosphere.

In the design of Tianmu Lake Service Area, the architects mainly use modern and neo-classical design methods. The architectural style suits the modern design of the Nanjing-Hangzhou Motorway and also attaches importance to cultural elements, as well as integrating into the surrounding environment. The service area is designed like a harbour. The vehicles on the motorway are akin to boats in need of a harbour to rest after a rough journey at sea.

The architectural design starts with focusing on general layout and use of space and different materials in order to successfully create a harbour-like atmosphere. The following measures were taken:

(1) A semicircular and rectangular design is used to form rich indoor and outdoor spaces (Figure 5.21). A semicircular square is naturally formed in front of the main entrance, welcoming visitors to the area (Figure 5.22).

(2) A continuous outdoor corridor is arranged along the outside of the main building, which ensures that people can move around conveniently and also connects the different functional areas together and consolidates the

Figure 5.21 Outdoor scene of the semicircular square in the service area.

Figure 5.22   Semicircular square and fountain.

building's design. Meanwhile, continuous columns, open-air corridors and a continuous framework integrated with the square also contribute to creating a neo-classical urban square and increase the sense of intimacy (Figure 5.23).

(3) The large outer doors and windows of the main building are made of double-glazed glass. This type of glass provides a broad view to the outside; it can also provide insulation, conserve energy and is an excellent soundproofing material.

(4) The facades of the building are made of white granite decorative sheets. The decorative effect is both classical and modern, resulting in a cozy and

Figure 5.23   Indoor view of the rest hall.

Figure 5.24 Embossments showing strong local characteristics.

comfortable environment. The decorative stone is also solid and durable and retains its appeal (Figure 5.24).

(5) The fast-food restaurant and the rest hall are characterized by arc-shaped interior space, which provides people with a sense of centripetal force and flow, and presents typical features of public structures used for transportation purposes.

The architects have also optimized the layout of the interior functions in the main building for the convenience of visitors. The dining and rest area are separated from the toilets and they are connected via a corridor. This means that the two main functional zones are both separated and connected but do not interfere with each other. In addition, considering the regional characteristics of where the building is situated and the overall design requirements of the Nanjing-Hangzhou Motorway, a small zone has been designed in the main building for displaying and selling famous local specialties in Liyang such as tea, chestnuts, bamboo ware and artwork. This zone is convenient for travellers and promotes the local culture.

Selection of landscape materials

When selecting vegetation for the service area, the soil properties and climatic conditions must be considered. The aim should be to improve the natural environment within the area and make an overall plan for the greening of the area's landscape.

Because the conditions for greening are poor in the area, tree species should be selected according to the designed landscape and local climatic and geographical features. These tree species should have the following characteristics: evergreen or long green period, do not require pruning, developed root systems, drought-enduring, capable of absorbing harmful gases and resilient to barren soil pollution, pests and diseases. Whilst native plants should account for the majority, superior foreign tree species can also be introduced and selected. By referring to the characteristics of natural vegetation, plants should be combined appropriately so that they blend into the vegetation close to Liyang and the landscapes along the route.

The climatic environment also affects the colour of landscape materials selected. For example, as the service area is close to the city of Liyang, which experiences heavy fog and low visibility, it is advisable to select bright outdoor landscape materials with a high contrast and saturation. This helps to ensure the creation of a uniform and attractive service area environment.

In conclusion, during the design of the Tianmu Lake Service Area, the architects have created a pleasant harbour-like atmosphere and provided comfortable service facilities for visitors.

Taihu Lake Service Area

The Taihu Lake Service Area is located on the west bank of Taihu Lake, adjoining the Lake amongst a vast landscape of lakes and mountains. It is the first service area on the boundary between Jiangsu and Zhejiang provinces on the Nanjing-Hangzhou Motorway as well as being the largest service area on the route. It also forms the climax of the 'pearl necklace' concept.

Overall landscape design

The Taihu Lake Service Area is composed of east and west parts, which are distributed on both sides of the motorway. The comprehensive service building is built in the larger east part, whereas the smaller west part only features a petrol station and a small parking area. Both parts are connected by a flyover. Most of the functions, such as shops, catering, accommodation, entertainment, rest areas, parking, large toilet area and the vehicle maintenance and refuelling facilities are located in the east part.

Motorway and parking area: In China, the large parking area of a service area is normally built beside the motorway. The service buildings are built behind the parking area, and the main service building can be seen directly from the motorway, which results in noise, dust, exhaust emissions and reinforced cement in the surrounding environment, contravening the 'people-oriented' design principle.

The Taihu Lake Service Area is designed to be surrounded by woods and vegetation, with the main service building located away from the motorway and separate from the parking area. The parking area includes separate parking for small and large vehicles, both immersed in trees and vegetation. The parking spaces for small vehicles are near the service building and the parking spaces for larger vehicles are further away. In the service area, ancient Chinese garden landscaping techniques are used throughout, yet these techniques are expressed in an entirely modern way. The design ensures that the landscape is both uniform and has its own characteristics. This kind of layout means that the noise from the motorway will not interfere with the peaceful environment in the service area. The green mountains and clear water of Taihu Lake make for a very picturesque landscape.

Petrol station: The petrol station is arranged at the exit of the service area for convenient departure and refuelling. It is surrounded by plants to keep it relatively separate.

Main service building and other subsidiary buildings: Jiangsu Province is located in the Yangtze River delta region, with a hot and sunny climate in summer and heavy rain in spring and autumn. The front of the service building therefore features a large overhang where vehicles can park and passengers can get out to enter the service building. Visitors do therefore not need to worry about the weather. There is also a convenient road to enable cars to drive directly to the main route.

The entrance foyer serves as a central zone that connects to various different functional areas, so it is very important to ensure its appropriate size and dimensions. In the catering area, separate sections for fast-food, an outdoor cafe and a small restaurant have been built, each with their own kitchens. The toilets are also an important part of service areas. In the Taihu Lake Service Area, the toilets are located in the lobby of the service building. After entering the lobby, visitors pass by enticing shops and leisure facilities before reaching the toilets. Alternatively, visitors can access the toilets directly from a dedicated side entrance. This layout has resulted in infinite business opportunities for the Taihu Lake Service Area. In addition, the design of the service area has made otherwise mundane activities more interesting. There are specialized shops to meet visitors' needs, selling car accessories, pharmaceutical products, food, groceries, local specialties, daily essentials, books and maps. This has created a busy and successful retail area.

Paving: There are two sorts of materials used for paving the ground of the square in the service area, namely asphalt concrete for the roads and cement concrete for the parking area. The divisions between each area are clear, and each has its own function.

The square in front of the service building is paved with fired granite slabs, surrounding and following the arc shape of the building. The fired granite slabs are light grey and dark grey, simple yet attractive and imposing.

Road signs and curb: The road signs in the Taihu Lake Service Area are made of bent pure aluminium plates and the markings on the signs are made from reflecting material so that they are suited to daytime and nighttime environments. The design is simple and modern.

The curbs at both edges of the motorway are required to be conducive to parking and resistant to damage, so they are made of processed granite. The edge of the curb on the side of the parking area is sloped in order to adapt to the compression of vehicle wheels. After the sloped area the curb is 5 cm thick, resulting in a solid and reliable surface.

Landscape design: The Taihu Lake Service Area is located next to the imposing scenery of Taihu Lake, providing the service area with a particularly favourable natural landscape. The landscape design took this as its starting point. The design aims to create an excellent environment within the service area and minimize the impact of noise from the motorway.

Taking full advantage of Taihu lake

According to the functional needs, the service area is divided into five subareas, namely, a parking area, service area, shopping area, refuelling area and a connecting bridge area. Between the areas, dense groups of trees and shrubs as well as grass are planted to separate, shade and adorn the environment or open up and enclose surrounding areas. Groups of *Trachycarpus fortunei* are arranged on both sides of the road in the service area. They create an attractive coastal landscape rarely seen in inland areas of Central China and effectively divide the boundary between each area and the road.

The natural topography is the basic skeleton for the landscape within the service area. The topography is used to achieve a series of design intents: (i) oOrganize the view; (ii) divide spaces; (iii) form the background scenery; and (iv) create landscapes.

It is worth noting that there are two glass wall fountains on the square in front of the comprehensive service building. Clear water flows down from the top of the glass wall, forming a thin water curtain that looks beautiful in the surrounding environment and that does not impede the line of sight.

Building design: The actual floor area of the buildings in the service area is 7500 m^2, with three floors above the ground and one underground parking area. The buildings include a rest hall, shops, catering facilities, a kitchen, large toilet area, tea house, coffee house, gym, meeting room, offices, guest rooms, and a roof garden. The building is the biggest comprehensive service center on the Nanjing-Hangzhou Motorway.

The most impressive aspect of the Taihu Lake Service Area is the bright and grand architectural design used (Figure 5.25). The idea behind the large silver-grey curved roof stems from the beauty of the surrounding water. Its natural geometrical features give the building an artistic appearance and the wide applications of modern building materials give the building a modern look. Curved roofs with such a scale and structure are rare in Jiangsu Province. It is difficult to position the 2700 m^2 of hyperboloid roof. The requirements for

Figure 5.25   Design of the main building in the Taihu Lake Service Area.

a waterproof, energy-conserving and low-cost surface with smooth connections and smooth curves make the design even more difficult. The designers did a thorough study and comparison before marking out different sections in order to solve the problem. The building is made of large span reinforced cement with its exterior covered in aluminium alloy roof plates. The reinforced cement structure is used to form the sloped roof; and the aluminium alloy roof plates are also used for this purpose, ensuring the successful design of the roof. Compared with a steel structure, this design has the following advantages: (i) lower cost; (ii) it forms two waterproof protective layers; and (iii) the structural layer is combined with the metal cover and energy-saving and thermal insulation materials are used, consequently energy consumption and construction difficulties are greatly reduced.

The comprehensive building is made of a frame structure, with its roof made of a reinforced cement hyperboloid plate. The structural system has been designed as rationally as possible, and the structural calculations used are accurate, safe and reliable. The roof is hyperboloid, so the structural system of the building is very complicated.

There are mainly the following difficult points during the structural design: first, it is difficult to select a structural calculation mode; secondly, the roof span is large and the building shape imposes strict restrictions on the height and cross section of the column; and thirdly, the roof curve is irregular, resulting in difficulties in determining the elevation and size of the member bar.

For this reason, corresponding technical measures are taken during the structural design. Based on the curved architectural model provided, the elevation and dimensions of the end points of each roof member bar are determined, reducing construction difficulties. In addition, the structural design of the foyer roof is difficult due to its large span and rooftop windows. After analysing and comparing many proposals, primary frame girders were adopted to suit the shape of the rooftop windows. Curved secondary girders have been employed, thereby reducing the dead load of the roof plates and increasing the rigidity of the building. Calculations show that both the level of reinforcement and the cross section of the primary frame girders are smaller than those used previously. This has reduced costs and is in line with architectural aesthetic requirements.

Other subsidiary buildings have been designed at low cost to fit in with the comprehensive building in terms of shape, style and colour.

During building design, means such as shape, material, colour, structural style, composition, images and letters may be used to enhance the atmosphere and reflect certain local characteristics, including history and culture. Furthermore, sculptures, wall paintings and symbolic combined landscapes can be used in important scenic spots to strengthen the cultural theme of the area. For example, the toll station on the main route of the Nanjing section of the Nanjing-Hangzhou Motorway is designed in a wing-like shape, signifying the nearby Nanjing Lukou International Airport. It also reflects the open-mindedness of the Jiangsu people, the challenges that confront them in the new century and their sense of longing for a bright future.

The Taihu Lake Service Area relies on the beautiful lakeside environment of Taihu Lake. The large triple-curved roof looks like a shell allowing visitors to use their imaginations. The service area is very modern yet also contains many historical and cultural elements. A giant rock has been placed on the grass within the parking area, on which the two characters for 'Taihu Lake' are carved in bright red. The rock is surrounded by a green landscape (Figure 5.26). The sculpture pool in the entrance hall depicts a typical Jiangsu and Zhejiang rural landscape and is an enticing attraction for young children (Figure 5.27). Plum blossom, orchid, bamboo and chrysanthemum paintings have been hung on both sides of the corridor on the second floor, and also paintings depicting ancient Chinese stories, including Zhou Chu (the hero in

Figure 5.26   Taihu Lake rock beside the parking area.

Figure 5.27   Attractive sculptures in the foyer.

the novel *A New Account of the Tales of the World*), Dongpo [or Su Si, a famous writer in the Northern Song Dynasty (960 – 1127)] and *Meeting on the Balcony* (a legendary story from *Liang Shanbo* and *Zhu Yingtai*). These paintings reflect China's extensive and profound culture as well as the rich cultural heritage of Jiangsu Province.

Conclusion

Motorway construction in China is currently undergoing rapid development. Whilst speeding up the construction of infrastructure, environmental protection must also be considered and guaranteed. Building areas should employ modern landscapes and combine functionality with the surrounding landscape in order to create an area that meets the demands of modern living and possesses modern features as well as its own character. Adopting this strategy creates an area that meets basic traffic needs and also functions as a perfect place for recreation. The area is aesthetically pleasing and features many cultural characteristics, providing a model for the future design of motorway buildings.

6 Environmental Landscape Design of Bridges and Structures

6.1 Essential characteristics of bridges

Bridges are built for public use due to people's changing living and working needs. Guided by scientific and aesthetic principles, bridges are man-made structures that are built after meticulous planning using available resources and engineering techniques. They are the product of a combination of science and engineering technology. Advanced bridge construction not only reflects the development of human society but also the wisdom and creativity of humanity. They are also a symbol of the times and act as historical monuments. Their essential characteristics include:

(1) Bridges are static and stay static in the same landscape or environment.
(2) Although bridges are three-dimensional, they are constructed as an extension of a one-directional road according to the traffic needs. The structure of bridges therefore focuses on one-dimensional longitudinal design and they are typical traversing structures.
(3) Bridges are made of all types of materials and constructed into many forms; all bridges have their own mechanical characteristics in order to suit their different structures and span.
(4) As aesthetic objects, bridges' components are visible, which inevitably means that they will have a symbolic effect or even become landmark pieces of architecture. Bridges have a long life span and are often built at key places of human activities, thus they can be considered to constitute an important part of the beauty of nature, gardens and cities.

The Environment and Landscape in Motorway Design, First Edition.
Qian Guochao, Tang Shuyu, Zhao Min and Jing Chun.
© 2014 China Communications Press. Published 2014 by John Wiley & Sons, Ltd.

6.2   Coordination between bridges and the environment

Coordination issues in bridge aesthetics include the coordination between the bridge and the environment (including buildings), coordination between structural systems, the relationship between the whole of the bridge and individual sections as well as between individual sections, shapes formed by component member bars, bridge form, colour and cultural connotations. In summary, the issues relate to coordination between the bridge and the environment and between its own parts.

Coordination between bridge form and the environment

Matching the bridge form with its environment

Beam bridges: Beam bridges are straight, sturdy, straightforward and imposing, creating a powerful effect as they extend outwards. They integrate well into the flat and spacious river landscape and modern city architecture. Beam bridges arc therefore an appropriate choice for flat regions and modern cities (Figure 6.1).

Arch bridges: The curved shape of arch bridges is gentle and soft, simple and imposing. They feature Chinese artistic characteristics and integrate both cultural and natural landscapes. Arch bridges are suitable for use in ancient cities and water landscapes rich in culture. These environments feature crisscrossing rivers and canals, tall ancient pagodas and ubiquitous

Figure 6.1   Beam bridge.

pavilions, verandas and temples. History has shown that pagodas, pavilions and verandas are inseparable from arch bridges. Their architectural forms consist mainly of arches and round roofs and domes, meaning that arch bridges blend well into such environments.

Arch bridges possess an intrinsic charm; they blend seamlessly into the surrounding undulating hills and their changing shape is very attractive. These bridges can combine function and form better than other bridge types and are more suitable for use in mountainous and gorge environments (Figure 6.2).

Rigid frame bridges: Rigid frame bridges are simple and imposing, and the direction of force transmission is very clear. They are frequently used in cases where a large clearance is needed under the bridge and the height is restricted. Slant-legged rigid frame bridges are especially suitable over V-shaped valleys or interchanges that only require a small span. Their span is small and there is a great degree of freedom in terms of shape design. It is therefore beneficial to calculate the dimensional proportions of each part according to the surrounding environment (Figure 6.3).

Suspension bridges and cable-stayed bridges: Suspension bridges and cable-stayed bridges are suitable for use over sea straits and in port cities. The beams, towers and main cables of suspension bridges are simple, featuring a combination of soft curves and bold straight lines. The longitudinal beam stretches far across the water, supported by the curve-shaped main cables.

Figure 6.2   Arch bridge over a gorge.

Figure 6.3 Slant-legged rigid frame bridge.

This creates a simple yet dynamic and modern look (Figure 6.4). The surface of cable-stayed bridges is very intricate and the girder line is succinct, continuous and smooth, creating a very strong traversing effect. Cable support towers are constructed to support the cables and form symbolic landmarks, forming an important component of the landscape (Figure 6.5). The cable support towers of these two types of bridges rise high into the sky, and the supported cables are curved and both bridges feature a very large span, meaning that the bridges blend well into the wide expanse of the sea, meandering coastline, protruding islands and rocky peaks. In port cities, there are large numbers of ships as well as cranes lifting and carrying cargo, and suspension bridges and cable-stayed bridges together with their cable support towers and supporting cables fit well into this environment (Figure 6.6).

Figure 6.4 Akashi–Kaikyo bridge.

Figure 6.5   Nanjing Yangtze river bridge.

Figure 6.6   Tsingma bridge, Hong Kong.

Geographical environment characteristics and bridge types

Different bridge types may be appropriate for different places and environments. What type of bridge should be selected to achieve a satisfactory outcome in terms of overall design? In the following different geographical environments, the appropriate choice of bridge will be explained.

Flat topography: Using beam bridges on flat topography can fully exploit their horizontal appeal and ensure unity with the surrounding environment. Beam bridges are stable and can therefore also increase people's sense of security.

Flat regions refer to areas with a flat expansive landscape, so bridges that are straight and continuous are most suitable for these environments, such as continuous bridges, continuous rigid frame bridges, continuous truss bridges and multi-arched bridges. However, flat regions often suffer from a monotonous landscape. If bridges are used to regulate the environment, then it is necessary to accentuate the effect that the bridge has on reorganizing the landscape. Large bridges such as half-through and through arch bridges, cable-stayed bridges and suspension bridges can be used (Figures 6.7 and 6.8). If the water surface is wide, the bridge deck is high or the main bridge has a long span or a clear convex longitudinal vertical curve, beam bridges or rigid frame bridges can accentuate the effect that the bridge has on the landscape.

Mountainous terrain: In mountainous environments with a distinctive layered landscape, continuous mountains form a green backdrop around the bridges. The dimensions and scale of a mountain mass are far larger than a bridge; meanwhile, the contours of mountains are irregular and make for an enticing and towering landscape. High cable-stayed bridges and suspension bridges are not very suitable for this kind of environment because they possess the same towering characteristics as mountains. In undulating mountainous environments, slant-legged rigid frame bridges or arch bridges are suitable when a bridge needs to be built over a deep V-shaped valley with small span, creating a sensation that the bridge is 'flying' over the landscape below. The visual impact of these two types of bridges is transmitted along the axes on both sides, which is just opposite to that of the mountain mass, leading to an effect of having been embraced by the mountains via their curved outlines;

Figure 6.7 **Japanese cable-stayed bridge.**

Figure 6.8   A cable-stayed bridge over a river.

this makes it easier for the bridge to integrate into the natural geographical environment. The Salginatobel Bridge in Switzerland (Figure 6.9) is a beautifully designed three-hinged arch bridge resembling a slant-legged rigid frame bridge. This bridge spans a steep gorge and blends very well into the geographical environment. Its white structure traverses an emerald green valley, making it appear even more stunning. If beam bridges with tall piers

Figure 6.9   Salginatobel bridge, Switzerland.

are chosen in this kind of environment, it may result in unfavourable visual effects, and thick piers must be used due to the height, which only exacerbates the visually clumsy appearance of the bridge. In the final 1999 issue of *Bridge Design & Engineering*, 30 world-famous bridge engineers, architects and scholars were consulted to determine the 'most beautiful bridges of the 20th century'; 16 bridges were subsequently nominated and the Salginatobel Bridge designed by R. Maillart was ranked first.

The Ruck-A-Chucky Bridge (Figure 6.10) spreads out its stay cables extending along the hillsides, guiding the axial force along the curved bridge deck. This means that there is no bending moment or shearing force produced in both upward and downward and left and right directions. The giant beautiful fan-shaped curvature is similar to a hyperboloid structure. The bridge design is conducted according to local conditions and has its own character. Though this type of bridge has not been used along the Nanjing-Hangzhou Motorway, it provides a good example that combines functionality and mechanics with the landscape.

If a bridge is to be built over a small river surrounded by gentle hills, meandering streams and distant mountains, then a beam bridge or multi-arch bridge with multi-span, large-span and low-beam may be chosen, as they are inconspicuous in the valley and fit in well with the surrounding environment.

Figure 6.10   Ruck-A-Chucky bridge.

The water flow of riverways in mountainous areas is generally small and its level varies sharply according to the seasons. When building a bridge over a seasonal wide shallow riverbed, more importance should be attached to the spatial form under the bridge and the proportions on and underneath it. If a multi-span continuous bridge is built, the arch pier must be buried deeply, close to the river bottom, it is therefore not possible to see the complete arch shape, giving one a sense of repression which can have an oppressive effect. If a suspension bridge or cable-stayed bridge is constructed in this type of terrain, the abutment under the bridge is very short, despite the fact that the cable support tower is imposing, resulting in an imbalance in the proportions of the sections above and below the bridge. When building a bridge on this type of terrain, attention should therefore be paid to adjusting measures to local conditions and selecting appropriate bridges. It is not a good idea to focus only on building new types of bridges and bridges with a large span.

Harbour regions: In the harbour environments, the landscape is expansive and the outlines on the horizon are low and simple, meaning that the visual impact is not as strong. In such an environment, it is easy to balance the landscape when building large bridges because the dimensions of the bridge and surrounding landscape can be kept uniform. Cable-related bridge types (suspension bridge, cable-stayed bridge or a combination of both) have an upward visual impact and large span, and employing them in this environment can maximize their allure and make them become a part of the landscape. For example, the Normandy Bridge (Figure 6.11) in France has a streamlined combined bridge deck and inverse Y-shaped towers that are simple, attractive, bright and stable and fit very well into the surrounding environment. Large half-through or through type arch bridges combined with accentuating measures can also form a beautiful landscape with the surrounding environment.

Figure 6.11   Normandy bridge, France.

Figure 6.12   Sydney harbour bridge, Australia.

For example, the steel truss arch bridge (Figure 6.12) in Sydney Harbour in Australia, 503 m in main span and 49 m in width, stretches over the harbour and complements the Sydney Opera House to form a magnificent landscape. Beam bridges with a horizontal visual effect can also easily become a part of the surrounding landscape because the bridge has a very similar shape to the surroundings.

Urban areas: Urban areas feature a large number of housing areas and public facilities, resulting in a built-up and irregular landscape. For this reason, small- and medium-scale bridges should constitute the main types of bridges selected. Urban bridge construction plays an important role in forming the city image and has a great impact on the surrounding environment. Urban bridges should therefore be designed based on the principle of protecting, utilizing, improving and creating the surrounding environment. For example, Tower Bridge (Figure 6.13) in London, UK connects the north and south districts of the city together. Two piers are erected in the middle of the river, on

Figure 6.13   Tower bridge, London, UK.

which enormous square towers are constructed. Between the two towers, there are two single-span bridges for different uses. The lower level is a movable bridge under which large ocean vessels can pass. The upper level is a permanent foot bridge. There are stairs within the towers, as well as other facilities, such as a museum, exhibition hall, shop and bar. Ascending to the top of the towers provides a spectacular view over the River Thames.

In order to ensure the bridge blends into the surrounding landscape, the style, dimensions and details of the bridge should be considered based on the importance, value and permanence of the neighbouring buildings:

(1)   The bridge type should complement the overall style of urban buildings, especially in famous cities with a rich historic and cultural heritage or in regions with an intact folk culture. For example, Chengyang Bridge in Sanjiang Dong Autonomous County, Guangxi Province features a unique regional style and reflects the traditions of the Dong minority's ancient culture.

(2)   The bridge size should complement the general layout of buildings planned in the city (Figure 6.14).

Methods to coordinate bridge form and the environment

There are three ways to coordinate the bridge and the surrounding environment: elimination (concealing the existence of the bridge), accentuation (highlighting the existence of the bridge) and integration (where the bridge and environment are essentially of the same style). When the existence of the bridge damages the appearance of the environment or landscape, the elimination method may be used to keep the bridge inconspicuous and concealed in the environment; when the bridge is intended to dominate the environment or act as a landmark, the accentuation method may be used; when the bridge is in between these two cases, that is neither elimination nor accentuation are required, the integration method should be used.

On a winding motorway, continuous white road markings clearly define the beautiful contours of the natural landscape. Sometimes people may not

Figure 6.14   Bridge in an ancient European town.

realize that they are driving across a bridge (Figure 6.15). Overhead curved bridges accentuate the curves of tall buildings and soften the high-rise landscape (Figure 6.16). In urban areas, bridges should be accentuated in order to achieve the harmony between the bridge and the modern landscape. The Forth Railway Bridge in Edinburgh, UK is located in the expansive landscape of the Forth River estuary. The bridge is large and imposing, so it forms a dynamic

Figure 6.15   Example of the elimination method.

Figure 6.16   Example of the accentuation method.

landscape with the surroundings and is also a unique local scenic attraction (Figure 6.17).

The Sunniberg Bridge (Figure 6.18) in Switzerland is also a low-tower cable-stayed bridge. The curved tower columns are an excellent complement to the beautiful mountain scenery. Piers with larger upper parts and smaller lower parts are needed for the pier top to restrict the torque, and the bridge's structure and function are integrated into its artistic design. By using the integration method, the bridge and the distant high mountains complement each other's beauty and add a poetic flavour to the natural scenery, therefore achieving a favourable aesthetic effect.

Figure 6.17   Forth railway bridge, Edinburgh, UK.

Figure 6.18   Sunniberg bridge, Switzerland.

Coordinating the proportions and dimensions of the bridge with the surroundings

Treatment of bridge proportions

The dimensions and proportions of a bridge refer to the relationship between different sections, including the proportions between the spatial parts (virtual) and the material parts (physical), concave and convex parts and higher and lower parts. The bridge and its parts should be given a suitable size and dimensions according to functional requirements, structural properties and aesthetic principles. Managing the proportional relationship between the bridge and its parts and between all parts is a key factor that influences the overall aesthetics of the bridge. Only when there is a good proportional relationship between the bridge and all details can the bridge achieve a uniform and harmonious effect.

When considering the proportions of the internal divisions of the bridge, spanning should be dealt with first because the proportional relationship between divided spans has a great influence on the overall effect of the bridge and its coordination with the surroundings. As for the span layout of bridge arches, the proportion between spans of many ancient Chinese three-arch and five-arch bridges is identical to the golden ratio. Figure 6.19 shows the Gongchen Bridge in Hangzhou. This bridge features appropriate proportions between all its parts, consequently it has achieved symmetry, stability, elegance, and coordination with the environment. Modern bridges focus on the aesthetic principle of simplicity and no longer feature golden ratio proportions. However, it is still advisable to use a suitable proportion to ensure rhythmic repetition in bridge architecture (Figure 6.20).

Figure 6.19 Gongchen bridge, Hangzhou.

Figure 6.20   A continuous beam with variable cross-section.

Treatment of bridge dimensions

Bridge dimensions refer to the proportional relationship between the bridge and certain specific standards commonly seen by people (e.g. the dimensions of railings, steps and vehicles). If both are uniform, the appearance of the bridge reflects its actual size, known as its natural dimensions or normal dimensions. However, there may be a discrepancy between the actual size of the bridge and its appearance. There may be two cases that result in the loss of normal dimensions: the first is when the actual dimensions are bigger, meaning that the bridge appears smaller than its actual size. This forms a special effect resulting in a free and abnormal feeling of intimacy (intimate dimensions). The second case occurs when the bridge appears larger than the actual dimensions (exaggerated dimensions). Urban bridges are never far from people and have a close relationship with us, therefore urban bridges should feature real, natural and normal dimensions. In scenic environments it is advisable to employ intimate dimensions in order to create a comfortable, pleasant and leisurely environment and enable people to appreciate the elegance and charm of the bridge. If accentuation measures are used to highlight the appearance of the bridge in the surroundings, then exaggerated dimensions may be used appropriately.

There are many elements involved in treating the dimensions of the bridges, and railings play an especially important role in the bridge's perceived size. The treatment of other details also has a major impact on the appearance of the dimensions. For example, the size of certain elements should never be increased or decreased during design, especially traditional floral ornaments and patterns on railings. This is because people have a fixed impression of the size of such objects, and therefore if they are enlarged excessively, people can no longer effectively estimate the dimensions of the bridge. Treatment of the density, thickness and degree of protrusion of decorative patterns is also required to ensure appropriate perceived dimensions. If they are too thick or thin, it will be detrimental to the overall uniformity due a loss of the normal dimensions. The treatment of dimensions varies according to different materials. The effects of looking at the bridge close up and from a distance should also be considered.

Complementing proportions and dimensions with the environment

When constructing a bridge, its proportions and dimensions should be fully considered with the environment in terms of layout plan and spatial combination. The bridge should fit into the environment and the environment should enhance the bridge's appearance. In landscape environments it is even more important that the bridge's proportions and dimensions complement and are in harmony with the environment (Figures 6.21 and 6.22). When considering the coordination between proportions and dimensions, it is worth paying special attention to the following three issues during design:

(1) The proportions and dimensions of the bridge should suit its mass and scale.

Figure 6.21 Proportional coordination.

Figure 6.22   Exaggeration in proportions.

(2)   Its mass and scale (especially the mass of the main bridge) should be in line with the scale of the river.

(3)   Its proportions and dimensions should be compatible with the square at either end of the bridge as well as the surrounding streets and buildings, in order to achieve a smooth, harmonious and rational layout and balanced space.

Coordination between bridges

Following the development of the transportation industry, more and more bridges are being built and are planned to be built over the same river or road. This therefore raises the inevitable question of how to coordinate different bridges along the same route.

Changing the scenery

Identical scenery whilst walking or travelling by boat can be monotonous and even make the journey seem longer. When changing the scenery it is essential to understand people's visual feelings. There are two visual requirements: first, a particular scene should remain for a certain period of time and there should be smooth intervals between different scenes; secondly, there should be uniformity during the change of scenery. This means that scenes should change based on a uniform layout and style. The scenes should also have a clear pattern and charm. Each scene should have a uniform style yet should also have its own characteristics and create a sense of familiarity as well as novelty.

Pleasant scenery calls for an orderly change in the landscape as well as uniformity amongst change. This orderly change and uniformity depends on the rhythms in the landscape. These rhythms can be both partial and general. Partial rhythms should be orderly whilst not monotonous; general rhythms refer to a landscape rhythm that is used repeatedly over different sections in order to achieve the greater uniformity of different landscapes and ensure a specific, abstract and common scene. The layout of scenes and the treatment of their appearance should be carried out according to the requirements of the scene, usually adopting the 'change–repetition–change' technique, so as to arouse interest. Estimating the visual impact is not an exact science but when adding change to the landscape, people are left with both a sense of familiarity and novelty.

Coordination principles between bridges

Bridges are a component of the landscape along the route and their layout and form should be dealt with such that they fit into the surrounding environment. The coordination between bridges depends on the rhythm in their design. The coordination principles between bridges are as follows:

(1) With regard to multiple bridges that traverse the same river or motorway and are located at a certain interval from each other, there should be diversity in form whilst assuring uniformity in overall style and mass (Figures 6.23 and 6.24). Design requires an overall layout before making adjustments. The series of stone arch bridges over the River Seine in France and arch bridges in Prague in the Czech Republic are good examples of seeking change in unity.

(2) Fully uniformity in shape should be assured for bridges that are very close or parallel to each other and old bridges that require widening, so as

Figure 6.23 Series of bridge examples.

Figure 6.24   Series of bridge examples.

to prevent visual disorder. If a bridge is designed with a strong rhythmic pattern, then two bridges next to each other increase the rhythmic effect (Figures 6.25 and 6.26).

Design idea

When designing a series of bridges, coordination between bridges and the environment should be considered as well as the relationship between the bridges themselves, according to landscape and tourism requirements.

Figure 6.25   Parallel bridges – coordination.

Figure 6.26   Parallel bridges – lack of coordination.

When touring along a river or road and passing many bridges over the same route, it is important to strengthen the bridge's systematicness and pay attention to the links between them. The conventional design concept for individual bridges must therefore be abandoned and a design concept for a series of bridges must be established.

Bridges' colours and materials and their connection with local characteristics

As the three key elements in visual art, shape, colour and material represent an organic whole. They are both interactive and mutually complementary. Shapes, which can be straight lines, curves, surfaces and solids, leave people with a visual impression. Colour and texture are the two properties of materials, and their hue, brightness, saturation, area and surface texture have a psychological impact on people, resulting in different feelings via the imagination. The colour and material of bridges have a great effect on people's feelings, so sufficient importance must be attached to them during design and they should be in harmony with the local characteristics of the place where the bridge is built.

Colour

Colour is reflected by shape. The decorative effect of colour is similar to the effect of clothes on people. The colour of the bridge should be selected according to actual circumstances, and consideration should also be given to whether the bridge is in harmony with the surrounding environment. The principle of colour harmony refers to a uniform relationship of both contrast and harmony, meaning that when choosing colours, a combination of different and similar colours are mixed together to create a harmonious effect. Human eyes are fond of a combination of hues but it is difficult to succeed in harmonizing more than three basic hues when using colours. During colour design, the simplest method is to arrange the hue of the primary colour over the largest area. It should have the lowest saturation of all the colours. The secondary colour should be arranged over the next largest area and have relatively high saturation. The area covered by the accent colour should be the smallest with the highest saturation. The bridge should have a primary colour and basic tone and contrasting colours are only suitable for use in small parts. Colours in large areas should be light and in small areas they should be darker in order to bring out the general tone. The primary colour should be light and can be used to accentuate detailed areas. A bright effect can only be achieved with light colours, and only by using diverse colours is it possible to overcome the dull nature of a construction. Warm and bright colours with a low saturation tend to be currently used.

As for the building colour, the most important factor is the selection of the primary hue. As an important factor in the colour of the environment, the colour of a bridge is used to define its appearance. The environmental colour of a bridge depends on how apparent the bridge is in the surrounding environment, namely whether it is integrated into the environment, attracts public attention, or is designed simply and plainly like the environment. Bridges are very practical constructions, so it is necessary to treat relevant parts with safety colours for vehicles and pedestrians. However, attention should be paid to eliminating strong light and dark contrasts, including on the bridge itself, and reducing glare due to light reflection. Different colours with a very strong contrast cannot be used on the same bridge even if they are safety colours, otherwise the effect will be very glaring. The above environmental colours are those which coordinate with the landscape and are compatible with the design and functional beauty of the construction. Coordination is achieved using a certain coordination method (any of the elimination, integration or accentuation methods). Safety colours are used for protective considerations and to fulfil the bridges' functional role. For example, railings, median barriers and lamp poles are painted with conspicuous colours.

It is favourable to select simple and light colours for the bridge. Small-area colour blocks are used to overcome the overall dullness of the structure and to complement and accentuate the space. In addition, it is a good idea to use bright colours or colours with a high reflectivity on the undersurface or piers in order to adjust the feeling of oppression under the bridge. (For example, the undersurface of flyovers is painted with bright colours.) The colour design of bridges is generally subject to the following process: based on the bridge shape completed, research the relationship between the bridge shape and different colours and ensure that there is uniformity of the whole structure, including the bridge, in order to achieve an attractive result. In other words, start with the determined bridge shape, select the environmental colour of the bridge by referring to the feeling, concrete and abstract associations of different colours, as well their attractiveness, and then determine the safety colours before finally adjusting the colours.

Attention should be paid to the following issues when considering the colours of bridges:

(1) Fully consider the influence of national cultural traditions and local customs, respect regional and ethnic colour preferences and give consideration to folk traditions and customs.

(2) By starting with the principle of coordination with the surrounding environmental colours, select the primary colour of the bridge, and consider the influence of the environmental colours upon the bridge (Figure 6.27). For example, for yellow, if surrounded by red, there are shades of green; and if encircled by green, there are shades of red. Another example is grey with equal tone, which produces visual differences in tone due to different base colours – on a dark base colour, it appears light, and vice versa. In addition, size illusion due to colour makes lighter objects appear bigger and darker objects appear smaller.

(3) Colors should highlight and accentuate shape, so as to enhance the shape of the bridge (Figure 6.28). For this purpose, the following expression methods may be used:

   (a) Accentuation. Use colours to accentuate the bridge shape and volume; for example, the edges of passenger foot-bridges can be treated with warm colours and the bridge itself with cool colours.

   (b) Enriching. For bridges that have an overly monotonous form, use colours to enrich their shape, that is use colour blocks in different

Figure 6.27   Hangzhou bay bridge.

Figure 6.28   Bay bridge, Japan.

shapes to enhance the bridge's form and add appeal. However, this effect is hard tomaster unless complicated comparisons have been conducted.

(c)   Summarizing. For the bridges that have a complicated form, pure colours can be used to summarize and unify the complicated form of the bridge, resulting in a pure, bright and imposing effect.

(d)   Division. Reduce the sense of heaviness of the bridge's form by means of colour division.

(e)   Contrasting. Use colours to contrast and accentuate the appearance of the main body and distinguish the primary and secondary areas of the structure.

(4)   During the colour treatment of bridges, colours must not only be in harmony and unity with each other, they must also be uniform with the features of the bridge. Because bridges are made of all types of materials, their colours and texture will inevitably be restricted and influenced by the building materials to a certain degree. As such, building materials must be selected in order to obtain good colour and textural effects.

Material texture

If the visual effects of a stone wall are compared with a concrete wall, the former looks natural and full of energy, while the latter looks grey and dull. From this comparison, we can see the role that material texture has to play. On many occasions, materials restrict the bridge type that can be chosen, and likewise, the bridge type determines the material that has to be used. Both materials and bridge types are restricted by environmental factors. If their relationship is managed effectively, they can complement each other well. If not, they will not blend well into the surroundings.

In current bridge construction, large quantities of steel and concrete are used, resulting in dull and monotonous colours. Together with the marks left over from construction, they look rough and unattractive. If these materials are used for the construction of landscape bridges, they will have a lasting impact on the environment. Landscape bridges are required to be harmonious in shape, bright in colour and natural in texture, therefore surface treatment must be carried out.

With regard to the material texture of bridges, the inherent appeal of structural materials should be utilized to the greatest extent; meanwhile, materials with local characteristics should be used. The natural appeal and contrast of materials should be used to achieve the construction of a simple and beautiful structure (Figures 6.29 and 6.30). Attention should be paid to the following issues when considering the material texture of bridges:

(1) The panels, girders and buttress of the bridge should be smooth, while the abutment should have a rough surface.
(2) When selecting materials for the bridge, the influence of natural conditions should be considered, and the materials should meet the requirements of the structure.
(3) In general, the visual effect is best when people are within a range of 25 m and they can see the material texture and patterns. It is important to consider the actual circumstances, and the degree of treatment of the surface texture should be determined by the possible distance of the viewer from the structure.
(4) When paving the sidewalk, suitable colours, textures and dimensions should be selected considering the local environment and climate.

Figure 6.29 Extradosed cable-stayed bridge, Japan.

Figure 6.30   Twenty-three-arch landscape bridge over the East Lake in Wuhan.

Other practical issues relating to the coordination between the bridge and the environment

(1) In the environmental design of building, bridges have for a long time been required to be the basic component of the landscape, while the surrounding buildings are classified into different grades, for example, temporary and permanent buildings. The bridge shape should blend into the surrounding natural features, while the surrounding buildings should base their style on the shape of the bridge such that the shape of the auxiliary buildings complements the bridge.

(2) The basic function of bridges is to extend roads. The vertical face and surface of the bridge should form a close and smooth connection with the road and focus on an overall effect that also includes the road.

(3) The shape of tower buildings and the materials used for their construction should be similar to the environmental surroundings. They should look like they are 'growing out of the ground' so that they naturally fit well into the environment. Different from the purpose of tower buildings, bridges carry people and vehicles and their shape is influenced by the regional ways of thinking and characteristics. Bridges also reflect the current styles and trends. Bridge form should therefore be designed in coordination with the environment and attention should be paid to the symbolic nature of the form chosen.

(4) Bridges should be combined with environmental greening whilst attention should also be paid to the contrast effect of the environment. The architectural form of a garden bridge should be adapted to gardening techniques, meaning that modern bridges should employ a modern garden design method, whilst classical bridges should adopt a classical gardening method. The shape of bridges should not only give full play to the features and merits of a riverside area, it should also make the surrounding green area part of the bridge landscape, integrating the

artificial elements with the natural ones, thereby creating an attractive and harmonious spatial environment.

(5)   The bridge should integrate with the river water to achieve a contrasting artistic effect.

Landscaping methods of bridges

Integration of the bridge shape with the surrounding environment can constitute an important element of an attractive landscape. In many ways, whether or not the bridge is distinguished usually depends on how the designer understands and uses the environmental conditions to combine landscapes (make landscapes according to local conditions) and borrow scenery. Importance should therefore be attached to this issue right from the overall layout through to the detailed treatment of the bridge. Bridges are one of key elements of landscape environments, whereas bridges themselves are the result of landscaping. Bridges are used to borrow, align and frame different landscapes. This enriches the appearance of a particular scene, and is a common tool used in the utilization of landscape resources.

Borrowed bridge landscape

Fine scenes near a bridge (in the field of view) are introduced to within the bridge landscape, which is known as a borrowed bridge landscape. The purpose of borrowing scenes is to enrich the composition of the landscape and make the scenes more characteristic and varied, so as to create a certain artistic feeling and enrich the landscape effect. There are many methods that can be used to borrow landscapes, such as 'borrowing from the distance, borrowing from nearby, borrowing from above, borrowing from below, and borrowing from a different time' (Cheng, 2009, p. 241). Borrowing from a distance generally refers to borrowing distant beautiful scenery and applying it on the bridge. For high areas, both borrowing from the distance and borrowing from below can be used, for example leaning against the bridge railings, one can admire the glistening light of waves and inverted reflections in the water below. Borrowing from the distance and borrowing from nearby are different only in terms of distance. Scenes from different seasons may be borrowed to express different artistic ideas. In landscape gardens, bridge landscapes are often used to express the harmony between emotions and the landscape. In such cases, bridge names and couplets are often used to adorn the surroundings and establish the theme.

Borrowing form is a common method used to compose landscapes. This means combining attractive near and distant architectural structures and natural scenes into the landscape (Figure 6.31). For example, the Yanhe River Bridge in front of Yan'an Pagoda Mountain (Figure 6.32) features harmonious proportions between all its parts and a rhythmical structure to its upper part.

Figure 6.31   Bridge over a motorway.

Figure 6.32   Yanhe River Bridge, Yan'an.

With the Baota Mountain and ancient pagoda in the background, the style of the bridge blends into the surroundings and its shape contrasts well with the landscape. This forms a vivid and moving landscape that symbolizes the revolutionary nature of Yan'an (an important region in China's Red Revolution). In this distinctive style, the stone arch bridge forms a unique scene with the pagoda, mountain and water and they contrast well with each other. All three elements are needed to achieve such an effect. The Yanhe River Bridge is in itself not that imposing but it is in harmony with the surrounding environment of the mountain, water and sky. As a result, the Yanhe River Bridge has been printed on various different stamps over the years.

Aligned bridge scene

An aligned bridge scene refers to a scene where there is a landscape(s) at one end or both ends of the bridge. The landscapes are opposite to and aligned with each other. This is a kind of scene borrowing (Figure 6.33). An aligned bridge scene may be arranged exactly ahead of the vehicle to create a clear sense of direction and perception of distance. This gives the road and bridge a unique character (Figure 6.34).

Figure 6.33   Zhuangyuan Bridge, Hunan.

Figure 6.34   A bridge in Beihai Park, Beijing.

Framed bridge scene

Bridge arches, cable supporting towers, portals, railings and columns often form a viewing frame. In the frame, an aligned scene may be arranged or a distant scene may be borrowed to produce contrasting levels of internal and external, dark and bright, close and distant, and artificial and natural scenes. Real scenes viewed through a scene frame give people an illusion of looking at a painting, adding artistic charm to the natural beauty. Standing before the fishing terrace of the Slender West Lake in Yangzhou, visitors are treated to a panoramic view of the Five-pavilion Bridge and White Pagoda when looking southwest through two circular doorways. Flexible use of the scene framing method on garden bridges can enrich the landscape and provide different scenes as people walk from one place to another.

Considerations in bridge landscaping:

(1) The axis and location of the bridge are usually determined by traffic needs. When its direction and location have been determined, the existing scenes around the bridge are relatively fixed in place. This means that bridge landscaping is quite passive. When landscaping can be performed, harmony between the bridge shape and the surrounding scenery should first be assured. Then scene borrowing, aligning and framing can be used accordingly.

(2) There are specific requirements for combining scenery when building squares at one end of a bridge, viewing platforms, stairways and bridge corridors. These are important locations for the design of scene combination and borrowing.

(3) During the design of railings and cable support towers, attention should be paid to the application of the scene framing method.

(4) The human field of view is about 160° in the horizontal direction, 50° upwards and 70° downwards. Within this field of view, the optical axis (main line of sight) is located at an angle of depression of approximately 15°. In the process of visual perception, the nearer the objective image is to the optical axis, the higher the definition. Objects at the perimeter of the field of view are easily neglected. If the point of sight is in motion, attention should be paid to the direction of the optical axis of the point of sight. Ideally, the optical axis should be exactly on the most attractive landscape; the direction of the optical axis depends on the direction people are walking in. Zigzagging bridges in gardens not only enhance the appearance of the bridge itself, they also consciously organize the field of view and the direction of the optical axis. However, some zigzagging bridges constantly change direction randomly, which can be monotonous for the visitor.

(5) Importance should be attached to the landscape design of bridges. The landscape design of bridges refers to combining bridge types,

traffic, the surrounding environment of the bridge location and local cultural and historical characteristics together in design. The idea is to improve the appearance of the bridge and its environment and exploit scenic resources. It should be carried out according to bridge landscape construction standards and requirements, the objectives of landscape exploitation, regional construction plans and requirements for environmental protection. If sufficient construction funds are available, the bridge may be used as a landscaping element and appropriate landscaping design may be carried out.

6.3 Analysis of examples of bridges along the Nanjing-Hangzhou motorway

Bridge landscape design concept

(1) Natural landscape. The Nanjing-Hangzhou Motorway is located next to many mountains, rivers and lakes. The motorway landscape should therefore accentuate the local natural environment and integrate into the surroundings. The bridge shape should be natural and blend in well with the mountainous and river landscape.

(2) Profound historic and cultural heritage. It is very important to explore the cultural heritage along the Nanjing-Hangzhou Motorway. This should be combined with the functions and landscape of the motorway, allowing people to experience the rich cultural heritage in the region without even realizing it. Of course, the bridge itself is a product of the times but it is much more than simply a reproduction of the past. It provides the surroundings with a cultural and historical relic that is testament to new construction methods.

(3) Traffic and leisure. Following an improvement in education levels, people begin to re-examine their surrounding buildings and expect them to better reflect their needs. The Donglushan Bridge is an important traffic facility and scenic spot along the route. Each visitor can experience the convenience of the bridge and stop to appreciate the surrounding landscape.

Bridge landscape design principles

When enhancing and detailing the design of bridge landscapes along the Nanjing-Hangzhou Motorway, the following several principles are employed:

(1) Symbol principle. The bridge landscape should fully reflect the scenic characteristics of the Nanjing-Hangzhou Motorway, inherit and continue the local historical traditions and create a bridge culture and environment with distinct modern characteristics by using modern

design concepts and techniques. The Donglushan Bridge is full of poetic character. Both the overall shape of the bridge itself and the layout of the railings and bridgehead reflect a famous saying by Confucius: 'The wise finds joy in water, the benevolent finds joy in mountains'. The Nanjing-Hangzhou Motorway also signifies the development of a new motorway after the construction of the Shanghai–Nanjing and Nanjing–Hangzhou Motorways. This will assist the expansion of the surrounding economy in the 21st century.

(2) Simplicity principle. The design should be straightforward and strengthen the coherence and uniformity of the landscape. There should be change amongst the uniformity, and contrasts should be accentuated in order to bring out the stylistic characteristics of the surroundings.

(3) Landmark principle. The space and environmental design should strengthen the landmark and identifiable characteristics of the surroundings, allowing people to experience the unique regional character and charm of the Nanjing-Hangzhou Motorway.

(4) 'Putting people first' principle. Human dimensions, human needs and human activity should be the fundamental considerations during design; all kinds of functions and service facilities should be provided.

Bridge landscape design style

Design has its own soul. It is also creative and central to achieving sustainability. With the aim of overall enhancement of motorway construction standards and increasing the quality of engineering, the design of the Nanjing-Hangzhou Motorway is intended to minimize the damage to and impact on the natural environment along the route and achieve harmony and unity between the motorway and cultural and natural landscapes. This has been achieved by adhering to innovative design, new ways of thinking, new concepts, new goals and new standards. The Nanjing-Hangzhou Motorway is therefore a brand new motorway that is sustainable and representative of Jiangsu Province. It integrates the concepts of nature, environmental protection, scenic landscapes and tourism.

As a product of modern industry, bridges feature a modern look in terms of their structure and materials. When determining the style of a bridge, the environmental scenery along the route should be depicted in a modern form by combining local culture and traditions. The style of the bridge should therefore both correspond to the local circumstances and fit with the design of the bridge's main body. The surrounding landscapes of the Nanjing-Hangzhou Motorway should therefore be of a simple and fresh modern style and reflect regional characteristics. This kind of concise style is in line with local people's tastes, values and cultural identity. It not only complements the bridges characteristics, it is also a sign of the Yangtze River Delta region looking towards the future and modernization. The style reflects the region's new attitude and industrious enthusiasm; it is a true manifestation of the local people's character in the new century and something that they are proud of.

Analysis of examples of landscape design

Overpass in the Donglushan Service Area

The landscape bridge in the Donglushan Service Area is a single-tower and single-cable, cable-stayed bridge without backstays. The cable slants down from the main tower to support the girder, forming a simple and stable triangular shape. Simple shapes such as these create a towering landscape which creates a modern atmosphere that is in line with the high speed vehicles. It forms a fast and streamlined rhythm and this type of bridge is one of the most popular types of bridge today. The bridge's visual appeal stems from the stiffening girder, main tower and main cable. The main tower creates an upward visual effect as it rises up into the sky, which echoes the tall and steep Donglushan Mountain (Figure 6.35). The back-stayed cables are arranged at one side of the bridge parallel to the deck. The tower is inclined and its enlarged base balances the tensile force at one side. Such a shape can eliminate the instability of the tower. The overall design of the bridge is L-shaped and its slim girder makes the tower very conspicuous. Compared with the girder, the tower seems as though it is a piece of commemorative architecture. The back-stayed cables are arranged only on one side and make the bridge seem even more breathtaking (Figure 6.36).

The area in the middle of the bridge is like a violin, providing people with a new cultural landscape. It contrasts with the main body of the bridge and adds colours to the bridge and the surrounding landscape (Figure 6.37).

Figure 6.35   Donglushan Mountain seen from a distance.

Figure 6.36   Panorama of overpass in the Donglushan Service Area.

Figure 6.37   Bridgehead - violin.

The railings form part of the overall shape of the bridge. The railings are shaped like musical notation and their design is straightforward, dexterous, continuous and smooth, which contrasts with and accentuates the overall beauty of the bridge. Their shape matches with the bridge type in style in order to enhance its appeal and continuous and smooth sense of rhythm, and avoid potentially dangerous features that can distract the attention of drivers, involving shapes that are disorderly, strange or distorted (Figure 6.38).

Lighting is used to accentuate the structural features and reflect the architectural style. The main tower and cable plane are cast with reflected light to

Figure 6.38   Railings – musical notation.

highlight the tall, straight and imposing main tower, while the cables appear dreamily thin and transparent in the night sky. This makes the overall shape of the bridge distinct and vivid at night. It contrasts with the mist-shrouded Donglushan Mountain and the night lighting of the service area nearby, forming a poetic and picturesque night landscape (Figure 6.39). The embedded anchor boxes and exposed joints are beautifully simple and smooth; and the joints have a studded appearance (Figure 6.40).

Figure 6.39   Lighting.

Figure 6.40   Anchor box of overpass in the Donglushan Service Area.

Figure 6.41   Greenery.

The green space fits in with the other landscape elements in the bridge environment. It is like a natural filter, removing dust and purifying the air. The trees provide shadow from the sun and lower the temperature. They also blend well into the rest spaces in the service area (Figure 6.41).

The paved landscape is an important part of the bridge landscape. The paving blocks on the sidewalk arouse people's sense of belonging and identity with nature. The paving style complements the artistic style of the bridge environment, creating a totally coordinated artistic space, akin to a melody which forms the basis for a piece of music (Figure 6.42).

Figure 6.42   Pavement.

Passenger foot-bridges and aqueducts

Due to the development of the transportation industry, more bridges are being built over the same route, which inevitably raises the question of how to coordinate these bridges. When driving along a route and admiring the surrounding landscape, people easily feel bored if the scenery is monotonous, and the journey can even seem longer. In such cases, the scenery should be changed. Changing the scenery should be considered from a visual

Figure 6.43   Aqueduct landscape.

perspective. A particular scene should remain for a certain distance and there should be intervals between different scenes but attention should be also paid to the consistency of the change. In a group of scenes, changes should be made whilst guaranteeing a consistent form and style. The change should also have a rhythm and pattern.

When driving along the Nanjing-Hangzhou Motorway, the first object that comes into sight is an arch in the distance (Figures 6.43 and 6.44). Driving closer, two bridges can be seen (Figure 6.45). Driving ahead further, there

Figure 6.44   Pedestrian bridge landscape.

Figure 6.45   Close-up view of Figure 6.44.

is another bridge (Figures 6.46 and 6.47). This is an example of change and unity, ensuring a rhythm and pattern to the change. With a rhythm of 'one bridge – two bridges – one bridge', interest is generated during repetition. Each landscape has its own characteristic. The style is consistent, creating a feeling of both familiarity and novelty. However, this change requires that two parts of the bridge structures do not produce a disorderly visual effect. This means

Figure 6.46 View of the motorway.

Figure 6.47 View of the motorway.

Figure 6.48   T-shaped pier with arc lines.

that the spanning layout of the structures and the shape of the substructure must be uniform, and the elevation of the superstructure should be essentially equal to that of the bottom of the girder. The arch bridge adopts a single curved T-shaped pier, which enriches the shape of the space below the bridge and produces a good visual effect (Figure 6.48).

Slant legged rigid framework and V-shaped frame continuous girder

The dimensions of all parts of the slant legged rigid frame bridge are small and have balanced proportions. It appears both delicate and forceful, and features rich and orderly lines. It appears to hang in the air, just like a floating ribbon, making the bridge appear very attractive. In addition, its large span depends on the horizontal thrust and prestress of the rigid frame, so it is very suitable to be used on branch overpasses along excavated sections. There are no piers in the central reservation, meaning that it directly traverses the motorway like a 'flying bird' (Figures 6.49 and 6.50). This bridge produces a downward visual effect, which is exactly opposite to the upward visual effect of the hills. A balanced visual effect is therefore easily obtained. Together with its curved outline, the bridge seamlessly blends into the geographical environment. When vehicles pass by a narrow excavated section, people often focus their sight on objects above them. The slant legged rigid frame akin to a 'flying bird' therefore creates a reinvigorating feeling and eliminates driver fatigue.

The V-shaped rigid frame continuous girder is a continuation of the pier and girder rigid frames. The V-shaped pier enhances the appearance of the pier itself and improves the spanning capability of the bridge. Due to

Figure 6.49   Panorama of slant legged rigid frame bridge.

Figure 6.50   Slant leg of slant legged rigid frame bridge.

a unique pier shape and low construction height, the bridge features the following characteristics: novel structure, intricate and attractive properties, appropriate stress level, uniformity, minimal vibration and a comfortable driver experience. This bridge type is therefore very suitable for flat regions with river networks in the Yangtze River Delta (Figures 6.51 and 6.52). When the central reservation of the motorway is narrow, the bridge can span directly across the motorway; when the central reservation is wide, a pier may be erected resulting in two openings underneath the bridge. The beautifully shaped bridge has been built near the Taihu Lake Service Area and can be seen by vehicles kilometres away, making for a pleasant landscape. The Taihu Lake

Figure 6.51   V-shaped rigid frame continuous girder.

Figure 6.52   V-shaped rigid frame continuous girder and greenery.

region is surrounded by mountains and water, forming a picture-perfect land-scape. The bridge is next to Taihu Lake; the backdrop of the bridge features distant mountains and it is surrounded by water. The scene fits in perfectly with the V-shaped girder, reflecting the 'nature, environmental protection, scenery and tourism' themes of the Nanjing-Hangzhou Motorway.

Tied-arch bridge over the Nanhe River

This is a bridge composed of two arches. The arch ribs include steel suspenders to support the deck, and the deck bears the tensile force, resulting in an arch bridge where the abutment is not subject to horizontal thrust. The wind braces strengthen the stability between arches. The main girder is thin and 130 m long. The whole tied-arch bridge is like a jade belt, or a rainbow, featuring 'soft lines and spaciousness' as well as a 'heavy and light, imposing and elegant' design. Its appeal stems from the graceful arcs, which result in a feeling of tension. In terms of its shape, the bridge integrates the beauty of cultural and natural landscapes and is a landmark structure. The red coating accentuates the appearance of the main body and is refreshing. Because the bridge is exposed and affected by such factors as sunlight, rain, wind, sand, ice and snow, weather-proof materials are selected that have a natural colour and texture. This also ensures an excellent colour and texture effect. The bridge type accommodates the contours of the road and blends into the surrounding topography, surface features and neighbouring landscapes. The beauty of the bridge's design and functions are therefore reflected and integrated perfectly. This type of tied-arch bridge is very suitable for flat areas (Figures 6.53 and 6.54).

Elevated bridges over the Beihe and Zhonghe Rivers

The Beihe and Zhonghe Rivers, located in the east and west of Liyang, are two of the region's major flood-causing rivers and feature different types of river traffic. These rivers pass through an ancient lake and marsh area between the Yixing–Liyang mountain area and the Ningzhen hilly area, where the terrain is low-lying with a ground elevation of 1.8–3.0 m and covered with deep soft soil. Many large fishponds are located between the Beihe and Zhonghe Rivers, meaning that the geological and engineering construction conditions are very unfavourable. Bridges built here should therefore avoid or reduce traversing north–south rivers, shorten the bridge area that passes over fishponds, and

Figure 6.53   Panorama of tied-arch bridge.

Figure 6.54   Red coating on tied-arch bridge.

increase the crossing angle with east–west rivers. Meanwhile, connection with the Zhenjiang–Liyang Motorway Hub should be considered to create favourable conditions (Figures 6.55 and 6.56).

Motorways occupy land and affect the ecological environment and daily lives of residents along the route. The occupation of land and the impact of the motorway on the surroundings are permanent. Motorway construction requires the sacrifice of land and surrounding environment. The destruction of the natural environment will often result in land desertification, water and soil loss, the greenhouse effect, ecological damage and the extinction of wild

Figure 6.55   Fishpond beside a bridge.

Figure 6.56   Bridges over the Beihe and Zhonghe Rivers.

animals. One important principle in motorway design is therefore to maintain a natural ecological balance as far as possible and minimize damage. In the past, people thought it was cheaper to build a road instead of a bridge. In order to reduce costs, they generally raised the ground instead of building a bridge. However, due to land protection policies, it is hard to acquire land and the environmental regulations are also very stringent. Analysis is therefore needed to determine whether building a road instead of a bridge really is cost-saving. This section is in a soft soil area. At the junction between the motorway and the bridge, the high-fill bridgehead slope often causes significant settlement of the subgrade. It also costs a significant amount to deal with the bump at the bridgehead arising from the differential settlement between the bridgehead road and the bridge. If a bridge is used instead of a road, the land below the bridge can be exploited and less land is occupied, meaning that the amount of filled earth and the impact on the environment is also reduced. To summarize, based on the natural, geological and hydrological conditions, an elevated bridge scheme is adopted by making comparisons from a societal, economic, technological and environmental point of view.

Continuous box girder

The continuous box girder creates a strong sense of movement as it extends left and right of the motorway. An equal-height girder is very suitable for medium- and small-span bridges on motorways without special requirements. The slenderness ratio of such a girder can vary significantly, meaning there is room for choice during design. The continuous box girder looks very light and harmonious and its slenderness can reduce the sense of oppression, creating an attractive and charming landscape. The stand-alone pier suits the changing and attractive contours of the motorway and also complements the box-shaped structure on the upper section (Figures 6.57 and 6.58).

Figure 6.57   Uniform section of continuous box girder.

Figure 6.58   Uniform section of continuous box girder.

7 Environmental Landscape Design along the Nanjing–Hangzhou Motorway

7.1 Design techniques

Landscape design along the route is based on an evaluation of environmental and visual scenery, with more than 10'hotspot' sections determined. By analysing various landscape elements, a technique is designed to enhance road patterns and features by using different forms, colours, textures and spatial senses. The position of each section and the corresponding analytical method is shown by the layout plan, while the detailed design is presented by a large proportional plan and 3D profile as well as pre-design and post-design perspective drawings (Figures 7.1 and 7.2). In order to improve the environmental landscape and reduce the gradient of the cut slope and embankment, native tree species are mainly used for green planting and plant landscaping, and a wetland is established to improve the drainage ditch, thus enhancing the visual appearance and relieving driver and passenger fatigue. Groups of trees are often needed at various sections in order to provide a strong visual impact. To achieve this visual effect, the trees must be planted densely.

7.2 Hotspot sections along the Nanjing-Hangzhou motorway

Key section I (LK1+700~LK2+900)

Figure 7.3 shows the layout plan of the design of key section I and the handling techniques, while Figure 7.4 presents the local details of plan design. Figures 7.5–7.12 are perspective drawings that display the landscape effects

The Environment and Landscape in Motorway Design, First Edition.
Qian Guochao, Tang Shuyu, Zhao Min and Jing Chun.
© 2014 China Communications Press. Published 2014 by John Wiley & Sons, Ltd.

Figure 7.1   Repeatedly structured planting method suitable for urban environment.

Figure 7.2   Natural trees dressing motorway like a country road.

before and after the design at four main sight points of key section I. It can be clearly seen from the figures that great changes have been made to the environmental landscape after the design, and the motorway totally integrates into the natural scene.

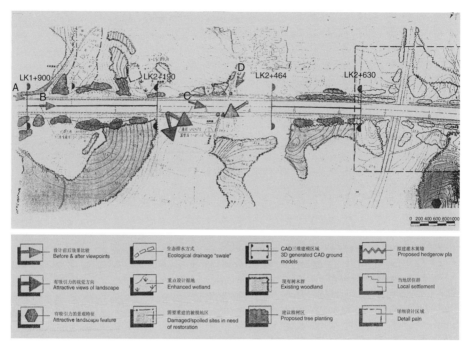

Figure 7.3   Layout plan of design of key section I.

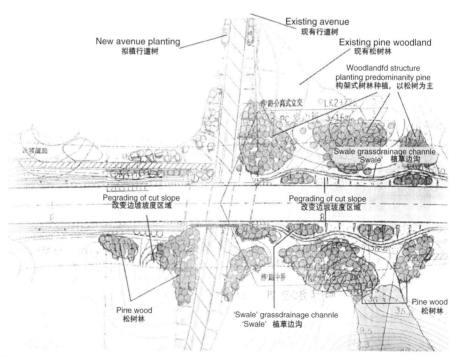

Figure 7.4   Local details of plan design of key section I.

Figure 7.5   Sight point A before design: slope protection with a simple lawn along the direction of the flyover.

Figure 7.6   Sight point A after design: improved slope protection with an upper part containing mixed planting species.

Figure 7.7   Sight point B before design: cut slope on both sides with mortar rubble for protection along the direction of the overpass.

Figure 7.8   Sight point B after design: reducing slope gradient and planting trees in a natural style.

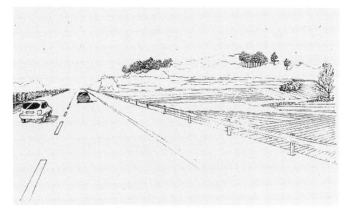

Figure 7.9   Sight point C before design: seen along the route, with passage in the foreground and hillside in the distance.

Figure 7.10   Sight point C after design: reducing the gradient of the distant mountain slope to integrate the newly planted trees with the existing trees; planting fruit trees and flowering shrubs at the nearby passage, indicating that the place is a design key point.

Figure 7.11   Sight point D before design:direct visual disturbance and prolonged noise pollution caused by the motorway, seen from a nearby village.

Figure 7.12   Sight point D after design: barrier-type planting eases the visual disturbance to the residents by the highway and improves the scenery along the route.

This road section is close to the toll station of the main line, with the design key points as follows:

(1)   Connecting the existing vegetation at both sides of the motorway with the pine woodland around the foot of the mountain.

(2)   Expanding the vegetation at the edge of the woodland to create new habitat for wildlife.

(3)   Adopting natural-style planting techniques and native tree species in order to enhance local characteristics and to build an attractive passage leading to the distinctive wing-shaped toll station of the main line.

Key section II (K40+20~K40+980)

Key section II is located at the back of a bridge that traverses a lake. The scenery is suitable for landscaping and greening (Figure 7.13), creating benefits for drivers, passengers and local residents. Along the motorway there is a small village, so attention should be paid to the following landscaping design issues:

(1) Expanding the bamboo zone to create a visual and noise barrier and ensure a good quality of life for local residents.
(2) Establishing a newly grown vegetation zone to improve and diversify the ecological environment.
(3) Guaranteeing that the landscape extends to the south of the motorway on account of the lake's natural charm, thereby promoting tourism and enhancing local characteristics.

Key section III (K50+200~K51+175) and key section IV (K51+175~K52+125)

Since key section III, key section IV and Donglushan scenery key section III are located on a rocky slope of Donglushan Mountain, it is difficult to carry out landscape design and technological work. Large quantities of rock and gravel are deposited at the roadside and planting zone but vegetation well covers the surrounding mountain body. This vegetation mainly includes evergreen

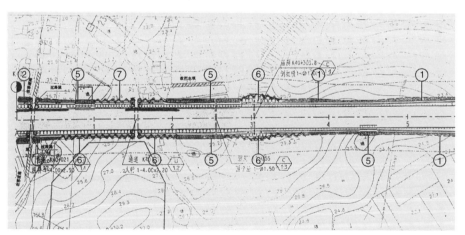

Figure 7.13   Layout plan of design of key section II: ①, structure planting; ②, rapid response structure plantingl; ⑤, water's edge structure planting; ⑥, hedgerow planting; ⑦, rapid response hedgerow planting.

secondary forest and dwarf bamboo as well as occasional tea plantations and other crops.

Landscape design (Figure 7.14) aims at:

(1) Establishing green slopes to adapt to the hillside low meadow.
(2) Mixed planting of wild flowers and grass to represent the diversity of native plants on the mountain slope.
(3) Treating the base surface of the slope with a unique approach to create a green belt of irregular weeds, shrubs and native plants and adapt to the conditions of the local scenery.
(4) Using irregular greenery, vines and climbing plants to decorate the horizontal concrete slope.
(5) Making the base surface of the slope gentle, applying a special tree planting technique to reduce the visual impact due to the slope height.

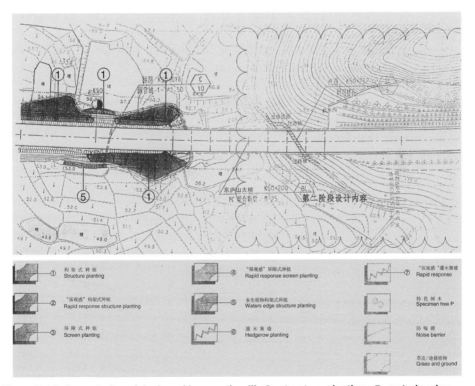

Figure 7.14   Layout plan of design of key section III: ①, structure planting; ⑤, water's edge structure planting.

Key section V (K60+600~K61+575)

Key section V features a beautiful environment and scenery, with a larger improvement space due to the existing vegetation zone and ideal topography (Figure 7.15). Landscape design aims at:

(1) Expanding the existing pine woodland, broadening the visual field and improving the ecological environment of the roadside lake by increasing aquatic vegetation.
(2) Trimming and narrowing the slope surface to build a flexible subgrade and reduce the impact of engineering design.
(3) Enhancing local characteristics and biodiversity by using native plant species.

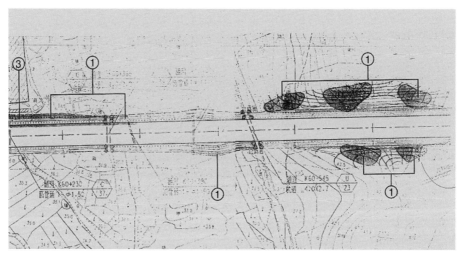

Figure 7.15   Landscape design plan of key section IV (K60+600–K61+575): ①, framed planting; ③, barrier-type planting.

8 Environmental Landscape Design of Drainage Systems

8.1 General requirements of motorway drainage systems

Surface water and ground water can be classified by different water sources. Surface water is mainly the surface runoff formed by precipitation (rain or snow), while ground water is mainly divided into perched water, phreatic water and interlayer water. A drainage system is composed of various drainage facilities and structures for intercepting, collecting, storing, conveying and discharging surface water or ground water. The aim of this system is to rapidly discharge the surface water and ground water, which is likely to endanger subgrade stability, to a place outside the subgrade using proper drainage facilities, and also to reduce the moisture content within the subgrade to within a predetermined range, so as to ensure subgrade strength, slope stability and driving safety.

For the drainage design of a highway subgrade, relevant meteorological data should be collected such as the precipitation frequency, cycle and precipitation process of the project site. For the regions with a cold climate, data about cold weather, freeze–thaw cycle and freeze depth should be collected. The topographical and geological conditions, soil type, soil thickness and groundwater conditions (water source, water amount) along the route should also be explored. Investigation and research should be carried out relating to various existing natural water systems or artificial drainage structures along the route, and integrated into the scope of drainage design. The drainage design of motorway subgrades and road surfaces should feature integrated planning and a rational layout. It should be combined with local irrigation systems for integrated design and to establish a variety of drainage facilities that form a perfect drainage system with complete functionality and a strong drainage capacity.

The drainage system design of the Nanjing-Hangzhou Motorway not only takes the drainage needs of the motorway into consideration but also strives

The Environment and Landscape in Motorway Design, First Edition.
Qian Guochao, Tang Shuyu, Zhao Min and Jing Chun.
© 2014 China Communications Press. Published 2014 by John Wiley & Sons, Ltd.

to achieve coordination between drainage system design and the overall landscape of the motorway in order to achieve consistency and harmony between project construction and environmental protection.

8.2 Consistency of design concepts and drainage system

The landscape design concepts of the Nanjing-Hangzhou Motorway can be condensed into two words: 'pearl necklace'. This means that the entire motorway looks like a necklace, while the interchanges, services areas and toll stations are regarded as the pearls of the necklace. This design concept serves as the soul of Nanjing-Hangzhou Motorway landscape design, based on which comprehensive and systematic landscape design is carried out in order to adjust measures to local conditions and achieve a landscape effect that is harmonious with nature and reflects the surrounding mountains and water.

With specific regard to the drainage system, it should not only meet the general motorway drainage requirements but also integrate into the overall landscape system of the motorway. According to the principle of reducing man-made scars on the landscape and integrating with nature, comprehensive consideration should be paid to various elements such as the road shoulder, slope, berm, ditch, stage for heaping debris, cutting, slope crest and intercepting ditch as well as the natural terrain along the route, in order to achieve harmony with the natural environment.

8.3 Innovation in the drainage system of the Nanjing-Hangzhou motorway

As the Nanjing-Hangzhou Motorway passes through upland and plain river networks, innovation in drainage system design has been achieved in the following aspects:

Drainage of pavement edge

According to the stipulations of the *Highway Drainage Design Specifications* (JTJ 018 – 96), the shoulder can be equipped with an intercepting water curb for collecting pavement water, and furnished with a slope chute at certain intervals for discharging the collected water. Since the longitudinal slope of flat areas is restricted by the spans of roads or rivers intersected by the route, the longitudinal slope of the general road section is gentle and the intercepting water curb of the shoulder should be arranged cautiously. Pavement water is mainly discharged to the outer edge of the pavement along the pavement cross slope. Arranging an intercepting water curb impedes the discharge of road surface water and forms a layer of water on the road, which constitutes a traffic safety hazard for high-speed vehicles. In addition, some of the road surface water flows to the bottom of the longitudinal slope along the edge of hard shoulder, and ponding of water occurs at the recesses of the pavement edge. This means that the asphalt concrete pavement is soaked with water for a long time, leading to loose parts and even pot holes. For this reason, an intercepting water curb of the shoulder should not be constructed, and a decentralized water-discharging method should be adopted. Additionally, protection measures should be taken to reduce the scouring and erosion of the pavement water on the soil shoulder and the subgrade slope.

The Nanjing-Hangzhou Motorway is made of asphalt concrete pavement, with the upper layer made of stone mastic asphalt mixture SMA-13, which is characterized by a sound resilience to rutting, cracking, skidding, aging and water. It also features good cement stability. The terrain along the Nanjing-Hangzhou Motorway is mainly flat and with small hills; the main route has excavated and filled sections. Based on the topographical features, for high-filled road sections, the traditional drainage design of constructing an intercepting water curb at the pavement edge is replaced with constructing a shallow-dish sump at the soil shoulder (Figures 8.1 and 8.2), while soil

Figure 8.1   View of shallow-dish drainage and slope greening.

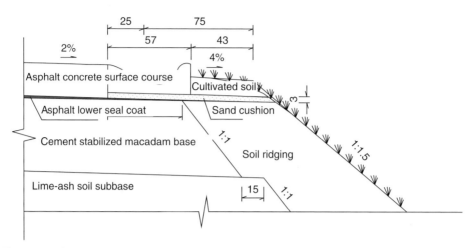

Figure 8.2   Schematic diagram of design of shallow-dish drainage (unit: cm).

ridging, grass planting and greening works is carried out at the outer edge. For low-filled road sections and excavated sections, due to a subgrade slope gradient of 1 : 6 or even an almost flat subgrade, the drainage system of the pavement edge is totally cancelled and the road surface water directly flows to the side ditch over the pavement cross slope via a gentle grass-covered green side slope. The interlayer water of the pavement structure directly flows into the green slope (Figures 8.3 and 8.4). The method of reducing slope gradient for low-filled road sections and excavated sections on the Nanjing-Hangzhou Motorway weakens the boundaries of the soil shoulder and the subgrade slope, and leaves space for greenery outside the shoulder. In addition to not using shoulder guardrails and constructing a buried ditch, flowers and trees of different height are planted on the slope outside the shoulder, creating a natural landscape effect.

Figure 8.3   View of overflow-type drainage and slope greening.

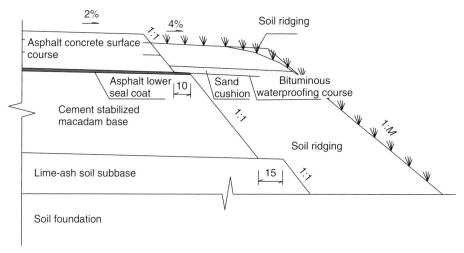

Figure 8.4   Schematic diagram of design of overflow-type drainage (unit: cm).

Drainage of central reservation

As the main ancillary facility of the motorway, the design of the central reservation increasingly reflects its importance in terms of functions, environmental protection, greening and landscaping. Central reservations can not only split two-way traffic and reserve space for laying communication lines; they can also guide the drivers' field of view after planting trees and vegetation, improving the driving environment. This has already become an important role of central reservations in addition to meeting motorway functions.

In general, central reservations include the following three types: flush type, shallow-dish type and recessed/raised type. For the flush central reservation, a masonry enclosure (or enclosure of rolled pavement materials) is adopted at the top, while internal pipes are not necessary, which can prevent rainwater from entering inside. Anti-glare facilities must be constructed separately. Since greening cannot be carried out, the flush central reservation type is generally no longer used. For the shallow-dish central reservation, an open-type all-greening scheme is adopted; the advantages of this type include greening and glare prevention. The flush and smooth curbs can prevent vehicle collision and bounce; the drawbacks include relatively thin filling and a low survival rate of planted trees. According to the *Highway Alignment Design Specifications* (JTG D20-2006), a recessed central reservation is suitable for central reservations with a width greater than 4.5 m, while a raised one is suitable for central reservations with a width less than or equal to 4.5 m. Based on comprehensive consideration, all-greening raised central reservations should be adopted for the Nanjing-Hangzhou Motorway.

For the all-greening raised central reservation, in order to prevent pavement pollution from rainwater, the filling height of the inner edge should be about 5 cm lower than the curb. Meanwhile, in order to ensure the survival rate of planted trees, a certain thickness of filling should be provided below the raised top surface of the central reservation, and communication lines and vertical blind drainage ditches should be buried below. The water seepage below the central reservation should flow to the macadam blind ditch at the bottom, and then discharged outside the subgrade slope via sumps and horizontal drainage pipes constructed at intervals of about 40 m. The side wall of the pavement structural layer within the central reservation should feature waterproof geotextiles or waterproof asphalt in order to avoid rainwater or irrigation water seeping inside the pavement structure and subgrade. The all-greening and raised central reservation used along the Nanjing-Hangzhou Motorway not only possesses the basic functions of a central reservation, but also reserves space for planting trees and greenery in order to reduce the erosion to the soil ridge on the top surface (Figure 8.5). Evergreen trees at a height of about 1.5 m should be planted at the central reservation, with low flowering shrubs planted at both sides. The resulting layered landscape acts as a good visual guide for drivers, while the green central reservation can also alleviate driver and passenger fatigue. See Table 8.1 for a comparison of central reservation schemes.

Drainage of subgrade side ditch

The purpose of the subgrade side ditch is to discharge the precipitation on the pavement and subgrade surface, preventing rainwater erosion to the subgrade and ensuring the stability of the pavement. Considering environmental landscape needs, the design of the subgrade side ditch should consider the drainage design of the side ditch, while more attention should be paid to a combination of side ditch design with landscaping and visual effects. Comprehensive analysis should be carried out based on geographical conditions, soil type, rainfall, slope form and other factors. It is essential to actively absorb design ideas and learn from the successful experiences of international and Chinese drainage works in order to reasonably determine the section form and geometric size of the side ditch. The side ditch design of the Nanjing-Hangzhou Motorway should mainly comply with the following principles: determine the section form and geometric size of the subgrade side ditch based on the catchment

Figure 8.5　View of greening on central reservations.

Table 8.1 Comparison of central reservation schemes.

Design schemes	Raised all-green central reservation	Shallow-dish all-green central reservation	Flush central reservation
Design patterns	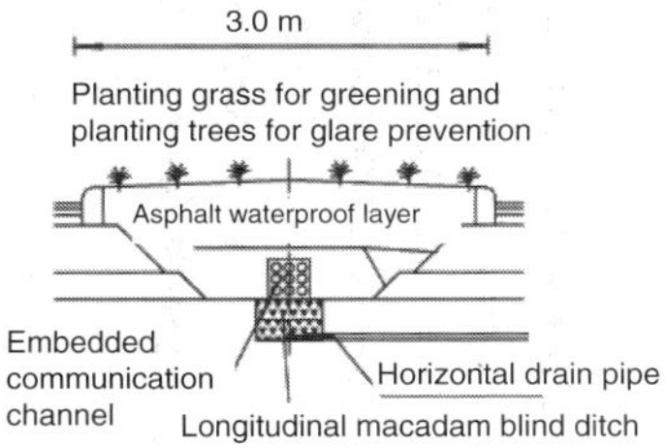 3.0 m Planting grass for greening and planting trees for glare prevention Asphalt waterproof layer Embedded communication channel Horizontal drain pipe Longitudinal macadam blind ditch	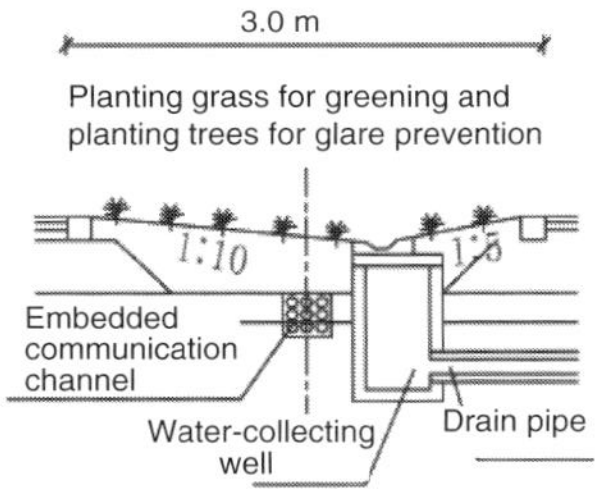 3.0 m Planting grass for greening and planting trees for glare prevention 1:10 1:5 Embedded communication channel Water-collecting well Drain pipe	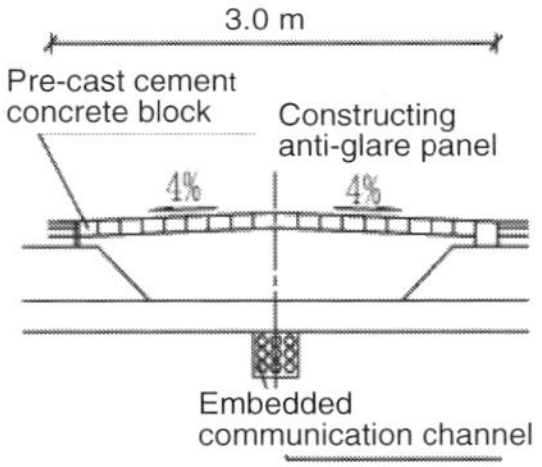 3.0 m Pre-cast cement concrete block Constructing anti-glare panel 4% 4% Embedded communication channel
Design key points	The central reservation is of the raised type, with soil filled and grass planted at the top surface, and raised curbs adopted. Some of the precipitation within the central reservation is discharged via the pavement, and part of the water seepage is discharged via a longitudinal macadam blind ditch and horizontal plastic drain pipe. Trees are planted along the route for glare prevention	The central reservation is of the shallow-dish type, with soil filled and grass planted at the top surface, and flush curbs adopted. The precipitation within the central reservation is discharged through a water-collecting well via a drain pipe. Trees are planted along the route for glare prevention	The central reservation is of the flush type, with masonry (or pavement materials) enclosure provided on the top surface, and with flush curbs adopted. The precipitation within the central reservation is discharged via the pavement. Anti-glare panels are constructed along the route

Table 8.2 *(Continued)*

Design schemes	Raised all-green central reservation	Shallow-dish all-green central reservation	Flush central reservation
Advantages	It highlights the road contours, provides guidance for drivers, achieves a sound drainage effect and complies with alignment design specifications. The greening and grass-planting are conducive to environmental protection	The drainage path is clear, with a good drainage effect; there are no curbs protruding from the pavement, to avoid bounce and overturning of small-sized vehicles due to the collision of wheels with the curb	Good drainage effect with no pavement pollution. Convenient construction with small engineering quantities. There are no curbs protruding from the pavement, therefore avoiding bounce and overturning of small-sized vehicles due to the collision of wheels with the curb
Disadvantages	The precipitation within the central reservation has a certain impact on driving when flowing over the pavement, and causes pavement pollution. The construction process is complex and there are stringent quality requirements	Complex construction process, difficult to control quality, low soil filling thickness, and low survival rate of planted trees. Difficult to construct drainage facilities for 3-m-wide central reservation	Cannot highlight the road contours well; relatively poor visual guidance function. The precipitation within the central reservation has a certain impact on driving when flowing over the pavement. Strict construction requirements otherwise a small amount of rainwater would infiltrate. Not conducive to grass planting, greening and environmental protection
Recommendation	Raised central reservations should be adopted for the Nanjing-Hangzhou Motorway after a comprehensive comparison		

area, longitudinal slope at the bottom of the ditch, materials and geographical conditions; achieve timely evacuation and the nearest diversion via a side ditch that should not be long; ensure consistency between the longitudinal slope of the side ditch and the road, and form relatively flat road sections in the local terrain. According to the results of calculations, the longitudinal slope of the side ditch should be appropriately reduced as long as no water collects in the side ditch; the side ditch elevation should be appropriately raised to avoid the backflow of farmland water; for sections where subgrade construction would cause significant damage to the terrain along the route, the subgrade side ditch should be adjusted according to actual circumstances.

According to the above principles, an arc-shaped side ditch has been adopted for the subgrade side slope at filling sections of the Nanjing-Hangzhou Motorway, with a width of 1.2 m at the top and 0.4 m at the bottom, and a depth of 0.4 m (Figures 8.6 and 8.7). In addition, in order to

Figure 8.6   View of greening of precast concrete ditch.

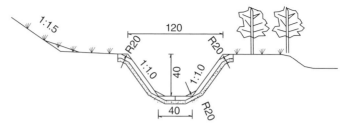

Figure 8.7   Schematic diagram of design of precast concrete ditch (unit: cm).

improve the internal quality and appearance of the side ditch, an advanced stamping process is also adopted for prefabricating, and the full-face of the arc-shaped side ditch consists of two symmetrical pieces, with 5 cm C15 pebble concrete adopted at the bottom for levelling. For sections with a short drainage length and small catchment area, a precast concrete side ditch should not be used on the basis that there is good plant protection on the slope, and a soil side ditch should formed in combination with site preparation (a little gravel can be spread in order to prevent erosion). Plants have been used for the greening of the soil side ditch and both sides, therefore naturally integrating the highway with the environment along the route (Figures 8.8 and 8.9).

Layout of cut ditches

In general, the upper layer of mulching soil on a mountain slope is loose, with a large permeability coefficient. The mulching soil gradually becomes denser from top to bottom, and the permeability coefficient becomes increasingly

Figure 8.8   Rendering of greening of soil side ditch.

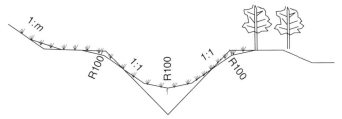

Figure 8.9   Schematic diagram of design of soil side ditch.

smaller. The lower layer is therefore aquitard relative to the upper layer. When perched water on the upper layer trickles down, only some of the water trickles down, and the rest accumulates at the interface of two layers, laterally flowing out of the side slope or the fractured areas. Therefore, in addition to hydraulic calculation, the section size of the intercepting ditch should be determined based on comprehensive consideration of vegetation, soil depth, soil type, soil porosity and other conditions. The longitudinal slope at the bottom of the intercepting ditch should be greater than 0.5%, in order to facilitate fast drainage of surface runoff. For road sections with a large catchment area, multiple rows of drainage ditches can be constructed based on actual requirements to prevent the cut slope from erosion. Accordingly, groundwater may escape from the rock and soil interface and between rock layers, so measures should be taken during the design to avoid slope collapse. In addition, the water outlet of the intercepting ditch should be specially designed based on field investigation to avoid the erosion of the subgrade slope at the cut-and-fill junction due to the mishandling of the outlet.

For the design of the Nanjing-Hangzhou Motorway, the flow rate of the intercepting ditch is determined according to the terrain and the precipitation at the project site. Since the amount of surface water flowing to the motorway is generally small (excluding the existing gullies, etc.), the size of the intercepting ditch at cut sections is initially determined as two kinds of rectangular sections of 40 cm×40 cm and 60 cm×60 cm. C20 cast-in-place cement concrete should be adopted at the bottom, and the side wall should be built with brick, with M10 cement mortar masonry used for plastering. The lateral platform, with a top width of not less than 1.0 m and a cross slope of 2% inward inclination, should be arranged inside the intercepting ditch in order to bring together surface water (Figures 8.10 and 8.11). For sections where the intercepting ditch of the partial embankment and the subgrade side ditch are constructed together, a 100 cm × 100 cm trapezoidal section should be adopted according to flow rate calculations, and precast concrete blocks should be used. The water outlet of the intercepting ditch should be arranged outwards as far as possible according to the terrain along the route in order to avoid damage to the subgrade slope and also reduce the impact on the visual landscape.

For some intercepting ditches with extreme difficulties in longitudinal drainage and small catchment areas, the traditional rectangular masonry

Figure 8.10   Rendering of greening of cut ditch.

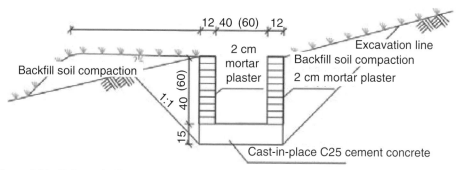

Figure 8.11   Schematic diagram of design of cut ditch (unit: cm).

chute practices should be changed, using buried tubular chutes for discharging the surface water accumulated on the cut slope surface to buried side ditches. The water is discharged through the intercepting ditch via a water-collecting well and buried pipe chute (30 cm double-wall corrugated pipe), then longitudinally discharged to drainage ditches or drainage structures along the route. Soil backfill and plant greening should be carried out at the slope surface with pipes buried below, to restore the ecosystem and protect the natural environment (Figures 8.12 and 8.13).

Mudstone outcrops exist at the Lanyoushan cut section on the Nanjing-Hangzhou Motorway. Considering such special circumstances, in order to prevent rainwater from entering as well as effectively reduce the impact of water seepage within the cut slope, inclined drainage holes should be constructed 10 m from the crest of the cut slope, more than 50 cm from the top of the mudstone surface for areas with mudstone and 30 cm from the foot of the slope. This design can ensure the moderate moisture content of

Figure 8.12   View of greening of buried chute.

cut slope surfaces to facilitate the growth of slope plants, and prevent large amounts of water seepage from accumulating within the slope and therefore affecting the overall stability of the cut slope.

Underground drainage

According to investigations of existing motorways, the pavement damage of low-filled sections, excavated sections and cut-and-fill junctions is much more severe than that of normal sections. Motorway designers should therefore attach great importance to the underground drainage of the above-mentioned areas. If the motorway passes through an aquifer or groundwater enrichment zone, measures should be taken depending on the actual circumstances to intercept, divert and discharge the groundwater of the aquifer in order to reduce the groundwater level or drain the groundwater within the slope. According to different functions and applications, underground drainage facilities can be divided into buried ditches (pipes), seepage wells and sewers.

During the design of Nanjing-Hangzhou Motorway, the practice of munic-ipal roads has been adopted, meaning that buried ditches instead of open ditches are used at excavated sections, and the circular catchment area is arranged at the top of the buried ditch, which is naturally integrated into the slope and covered with soil and plants. The rainwater is collected through an inspection well into a buried ditch and discharged outside the subgrade, so that the entire cross section is blended into one harmonious whole, which is natural and smooth. In addition, roadside crash barriers are not used at excavated sections in order to further widen the drivers' vision and ensure driving safety (Figures 8.14 and 8.15). In the original design plans, open ditch with mortar rubble protection is constructed at the drainage side ditch, the upper part with a width of 1.8 m and depth of 0.6 m, and the bottom part with a width of 0.6 m; the gradient of the inner and outer sides of the side

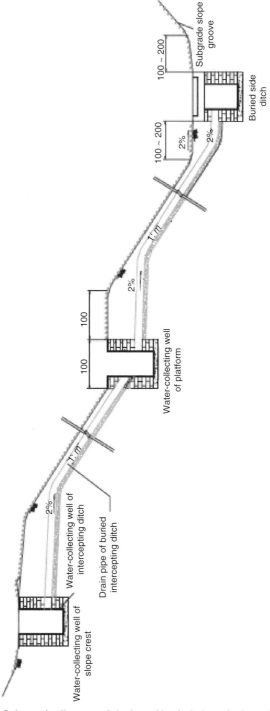

Figure 8.13 Schematic diagram of design of buried chute (unit: cm).

Figure 8.14   View of greening of buried ditch.

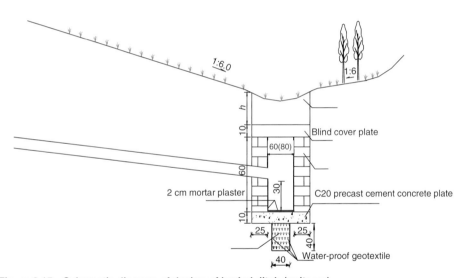

Figure 8.15   Schematic diagram of design of buried ditch (unit: cm).

ditch is 1:1. In line with landscape design and considering that excavated sections are usually relatively short, the longitudinal slope of the main line is relatively large, and the regional catchment area is relatively small and taking other factors into account, a buried side ditch instead of an open ditch is constructed at excavated sections. For the general excavated sections, a dish-shaped sink is arranged at the top of the buried side ditch. Precast cement concrete plates are placed at the bottom of the buried side ditch, the side wall is built with bricks and precast concrete covers are arranged at the top of buried side ditch. In order to collect the pavement water into the buried ditch via the dish-shaped sink at the top, the water-collecting well shall be longitudinally placed at intervals of 30–40 m along the buried side ditch, and an open cover plate with a weep hole should be constructed at the top of the well. Moreover, in order to prevent erosion of the subgrade due to groundwater at the outer edge of the side ditch, a macadam blind ditch should be arranged at the bottom of buried ditches on water seepage sections, which can effectively stabilize the water level. The top of the buried ditch should be trimmed into a circular shape and planted with grass for ecological protection.

Drainage of interchange area

Interchange areas occupy a large area of land and the speed of vehicles is relatively slow; they are an important scenic area. The drainage system of this area should feature appropriate landscaping that suits the local terrain.

(1) For an interchange with large ponds or surrounded by lakes, rivers and large areas of water, the water system should be adjusted to establish visually interrelated, well-spaced and consistently changing water-based green landscapes. Terrain renovation should comply with the original terrain and relief, in order to achieve an undulating effect and ensure consistency with the trend of the surrounding terrain. If the interchange is surrounded by a flat landscape, a slightly undulating terrain should be created to coordinate with the features of the surrounding plain water network. The so-formed landscape should integrate with the surrounding environment of the motorway and achieve a natural and ecological effect.

(2) If the interchange is near to urban areas, connection and contact with the urban drainage network should be ensured; there should be cooperation with the urban greening management department to establish a unified and coordinated greening arrangement and to avoid conflicts.

The drainage systems of all the interchanges on the Nanjing-Hangzhou Motorway are wetland treatment systems instead of conventional drainage ditches; this practice is a major innovation in interchange drainage of motorways and sets an example for the drainage works of future motorways. The specific procedures are as follows: cancel the internal side ditch of the interchange area; at shallow-excavated and low-filled road sections and on the inner side of the interchange ramp, appropriately reduce the gradient of the slope or reduce if possible in order to make the embankment slope or cut slope consistent with the original terrain or naturally integrate with the original terrain and avoid unattractive traces of artificial excavation or filling; and the steep slope and gentle slope can be combined depending on the

Figure 8.16 View of greening and drainage of interchange area.

slope rate of the terrain. Within the interchange area do not install a regular drainage side ditch in principle and combine environmental remediation with overall landscape design; based on the natural terrain or slight modification of the terrain, collect the rainwater into the wetland and existing ditches of the interchange area in order to address the drainage problems and improve the landscape effect within the interchange area (Figure 8.16). The creation of a large area of wetland and the weakening of scars on the landscape ensure that the man-made interchange is at one with nature.

Drainage of service areas and toll stations

The drainage system design of service areas should take the characteristics of the services into full account. The drainage system should fully consider the original terrain and relief and try not to destroy the original water system; avoid significant changes in the runoff mechanism of surface water, and avoid damage to the original ecological environment due to sewage; and buried sewage treatment equipment must be constructed at various management and service facilities that meet sewage treatment requirements. No kind of domestic or industrial sewage should be directly discharged into farmlands, rivers or lakes. After qualified purification treatment, the sewage can be discharged to wetlands or used for greening conservation of the motorway. During motorway construction in Jiangsu Province, Tianmu Lake Service Area was the first service area to adopt advanced water reuse technology for domestic wastewater treatment, realizing the reuse of sewage, which can save water and also reflect the environmental protection features of the motorway (Figure 8.17). For service areas near water, a trestle bridge from the main building to the water surface can be built, creating a waterside platform for people to enjoy the beautiful scenery around the lake (Figure 8.18).

Figure 8.17   Fountain at Tianmu Lake Service Area.

Figure 8.18   Waterside trestle bridge at the Donglushan Service Area.

Drainage of open spaces outside the ditch

Drainage systems outside the ditch should be designed based on the terrain, and coordinated with the local environment along the route. There exists a variety of situations as follows:

(1)   In the case of ponds, lakes or rivers near the outer side of the ditch, waterside plants should be planted according to local conditions, such as willow and aquatic plants (*Iris tectorum*, *Acorus calamus*, etc.), so as to restore the original scenery and represent the characteristics of a water village (Figure 8.19).

Figure 8.19   View of water seepage at the outer side of a side ditch.

(2)   When there is beautiful scenery around the outer side of the ditch and no dense residential areas, low shrubs and turf should be planted to integrate with the surrounding scenery and hide the side ditch in the green environment.

(3)   In the case of adjoining residential areas and schools or unsightly surroundings, tall trees should be planted in order to achieve noise and dust prevention and create a better landscape by mutually matching the colours, levels and species of trees.

(4)   In the case of surrounding stretches of vegetation and woods or a picturesque landscape, the land should be expanded at sloping hillsides or mountains in order to integrate the woods into the motorway. For example, for motorways surrounded by a large area of bamboo grove, bamboo can be planted at the side of the motorway. However, the terrain should vary according to the original terrain in order to make visitors feel that the motorway is natural, rich and colourful.

Conclusions

For the drainage system of motorways, focusing on an integrated design featuring environmental protection can reduce damage to original ecosystems, avoid or reduce water damage during construction and operation of the motorway and further improve the motorway's performance. This kind of design can also avoid and reduce environmental pollution and damage during construction and operation of the motorway, minimize damage to the natural environment along the route and achieve uniformity with the cultural and natural landscape.

9 Ecological Evaluation of Motorway Green Space Systems

The construction of motorways can greatly promote regional economic development but inevitably causes various negative impacts on the local natural environment, such as air pollution caused by automobile exhaust emissions, interference with the local ecosystem due to building/construction, and soil erosion. Green space systems on both sides of the motorway can improve the surrounding natural environment of the motorway due to their unique ecological value. They are therefore attracting more and more attention. During current green space construction, people have primarily focused on the landscape aesthetic value of green land but have to differing degrees ignored how to maximize its ecological value; therefore it is of particular importance to scientifically assess its ecological function and ecological value. Currently, the ecological evaluation of motorway green spaces is mainly focused on comments on the concept system, and there is a lack of concrete practice. Ecological theories are therefore adopted in this chapter to evaluate some of the green spaces along the Nanjing-Hangzhou Motorway.

9.1 Green space system of the Luojiabian Interchange Area

Main methods and coverage of ecological evaluation

The green space system of a motorway is regarded as an artificial ecosystem. Compared with a natural ecosystem, people pay more attention to its particular ecological service value. Nowadays, there is increasing research and practice in ecological engineering. The artificial ecosystem built by ecological engineering should first have the characteristics of a natural ecosystem,

The Environment and Landscape in Motorway Design, First Edition.
Qian Guochao, Tang Shuyu, Zhao Min and Jing Chun.
© 2014 China Communications Press. Published 2014 by John Wiley & Sons, Ltd.

that is substance circulation and energy flow as well as being self-sustaining, self- regulating and open. The construction of such an ecosystem should follow the basic guiding principle of 'from nature and superior to nature'. To be specific, an original stable natural ecosystem results from long-term adaptation to specific habitats, and a study on its composition can be used to guide the construction of artificial ecosystems, namely 'from nature'; on the other hand, artificial ecosystems are required to be strengthened relative to natural ecosystems in some functions, namely 'superior to nature'. Green space systems on motorways should therefore follow such general principles, and comprehensive consideration should be given to population and community levels, ecosystem levels and landscape levels in order to obtain scientific and accurate ecological evaluation.

The evaluation on population and community levels involves species composition, community structure and species diversity. The targets of evaluation on ecosystem levels mainly include productivity levels, species flow, energy flow and value stream, and the evaluation on landscape levels mainly covers landscape structure, landscape pattern and landscape diversity, with a detailed description given in the following.

Ecological evaluation on population and community levels

Species composition: Species composition in green space systems on motorways plays a decisive role in terms of ecological functions. Species selection should follow the following principles:

(1)   Dominated by native species. Native species adapt to specific habitats in the region (including climate, soil, etc.), which is conducive to individual survival and reproduction as well as the health of the entire system and ecological benefits. In addition, compared with those introduced exotic species, native species can decrease the risk of biological invasion. Invasive species have become a worldwide problem, causing very serious damage in many areas, such as *Eupatorium adenophorum* and *Alternanthera philoxeroides* introduced to China, and Canadian *Solidago virgaurea* introduced to Nanjing, China.

(2)   Dominated by broadleaf species with a larger leaf area index, higher productivity levels and strong pollution absorption and adsorption capacity. The special environment of the motorway determines this important principle of species selection. Since a large number of vehicle exhaust emissions lead to environment pollution, one of a green space system's main functions is to reduce or even eliminate pollutants and improve environmental quality. Species with a strong carbon assimilation capacity should therefore be selected. In this regard, broadleaf species are better than coniferous species, and species with a larger leaf area index are better than species with a smaller leaf area index.

(3)   Dominated by self-sustaining species. This is an important aspect for evaluating the health of a green space system. The stronger stress

resistance ability the species have, the stronger capacity the green space system has to resist disturbance. Since the habitat around the highway is unfavourable to most plants, species with strong stress resistance not only improve the stability of the green space system but can also significantly reduce conservation costs.

Community structure: Horizontal and vertical structures of the community as well as seasonal aspects have important impacts on the green space system. The proper layout of communities can help to reduce pollutants and lower noise. Examples include increasing high-intensity arrangement and community canopy density in heavy traffic areas. Attention should be paid to the vertical structure of the community during the construction of the green space system, such as the proper distribution of trees, shrubs and grasses, as well as the proper distribution of evergreen and deciduous trees. Compared with a homogeneous life-form, the above distributions not only increase the production capacity of the vegetation and improve the stress resistance and stability of the system but can also play an important role in landscape aesthetics. Currently, during the construction of ecosystems such as artificial forests, more attention is being increasingly paid to a proper community structure. For example, when building protective forests, the structure has begun to change from homogeneous species and unitary forests to multiple species and multiple layers of vegetation.

Species diversity: According to research results, species diversity is of great significance to ecosystems. Chinese and international research has shown that an ecosystem composed of only one or a few species has a relatively poor capacity to resist disturbance and disease, and has defects in terms of ecosystem functions. For instance, the outbreak of *Dendrolimus punctatus* Walker (commonly known as the Masson pine moth) in artificial *Pinus massoniana* (Chinese red pine) forests in China is in large part due to the existence of unitary forests. For zonal vegetation, due to the stable community composition and species diversity, higher biodiversity is not necessarily better because higher species diversity may reduce the system stability. In general, the species diversity of an artificial system is comparable with that of local natural ecosystem.

Evaluation of ecosystem level

The evaluation of the ecosystem level mainly focuses on the ecosystem functions, which include the following aspects: biological production, substance circulation, energy flow, species flow, information flow, value stream and decomposition.

One of the most important functions of an ecosystem is substance production. Plants absorb carbon dioxide through photosynthesis and release oxygen, and absorb and adsorb some of the air pollutants in this process. Meanwhile, a large amount of water vapour is released through transpiration, which can significantly improve the regional microclimate. In general, stronger transpiration indicates higher productivity. Substance production capacity is the basis of an ecosystem's service value. As for the green space system of the motorway, its ecosystem service value lies in its ability to improve the environment and reduce pollution. The main measurement standard of substance production capacity is Net Primary Productivity (NPP), that is, the total amount of carbon contained by vegetation over a unit time and unit area minus the carbon released for maintaining autotrophic respiration. NPP refers to the total energy which can be provided to other organisms (including humans), and is used as a quantitative indicator of an ecosystem function and ecological carrying capacity. NPP has now become a focal point of ecological research.

Species flow refers to the temporal and spatial variation state of a species community within an ecosystem or between ecosystems. It expands and strengthens exchange and links between different ecosystems, and improves the function of ecosystem services. The impacts of species flow on ecosystems include:

(1) Impacts brought about by increase and decrease of species. For example, on an island in the Pacific, due to the introduction of feeding moose, the community composition rapidly changed from hardwood forest to spruce forest.

(2) Invasive species alter ecological processes due to the use of resources. For example, the nitrogen-fixing plant flame tree has invaded Hawaii, making the nitrogen content in the local soil increase, providing fertile soil for the invasion of new species.

(3) Decomposition and its speed rate due to species loss and vacancy.

(4) The indirect impact of a change in ecological processes due to the impact of disturbance and stress mechanisms.

Substance circulation, involving various elements, media and chemical reactions, is a very complex process. It includes water circulation (the core part) and the circulation of mineral elements such as nitrogen, phosphorus and sulphur. As for the movement and transformation of toxic substances and radioactive substances, the absolute amount is small but it is closely related to human life. In green space systems on motorways, the movement and transformation of toxic substances contained in automobile exhaust emissions should be given close attention. Substances are the carriers of energy, while energy flow and substance circulation are closely linked. Upon the fixation of solar energy by plants, the flow in the ecosystem is one-way and gradually decreasing, with the food chain acting as the main flow channel. The energy flow rate in ecosystems is related to the stability of biomes and their capacity to resist disturbance. Due to a decline in productivity levels, a system suffering from relatively intense disturbance may result in a decrease of the energy flow rate of the entire system.

Information about ecosystems is characterized by diversity, complexities in communication, a large storage capacity and steadily increasing amount. Research on ecosystem information involves research on the generation, acquisition, transmission, processing, reproduction and invalidation of such information. According to different research objects, ecosystem information can be divided into information links between sunshine and plants, information links between plants, information flow between plants and animals and information between animals.

In ecosystems, the value flow is often linked to the energy flow. In 1981, H.T. Odum put forward a concept of energy value. Energy value means the value of another type of energy contained in the flowing or stored energy. In energy value analysis, the energy value is taken as a reference, and different varieties of energy which are not comparable can be compared after converting to the same standard. Solar energy value is often taken as the standard. In complex ecosystems, value and energy value flow in the opposite direction; for instance, exporting raw materials in exchange for money. Since ecosystems are a long-term natural product with a high energy value, people should first develop science and technology and improve resource utilization efficiency during economic activity.

Resource decomposition is of great significance for maintaining ecosystem functions and ecosystem services. The decomposition of dead individual organisms and degradation of pollutants are essential parts of the ecosystem. In some ecological restoration projects, decontamination is achieved by strengthening the decomposition role of the ecosystem.

Landscape evaluation

In current landscape ecology research, the following major landscape indices are adopted to describe and measure landscape structure and composition.

Shape coefficient: $D = P/2\sqrt{\pi A}$ referring to the ratio of the patch perimeter P to the circumference of the circle with the same area A to the patch.

Fractal index (fractal dimension): $D_f = 21(P/4)/1nA$, where the perimeter of the patch is P and the area is A.

Connectivity index: $CI = \sum_{\substack{j \neq k}}^{n} C_{ijk}/n_i(n_i - 1)/2$, with C_{ijk} referring to the connectivity between patch j and k (0 indicating disconnected, 1 indicating connected) and n_i refers to the number of patches.

Landscape fragmentation: $C = \sum_{i=1}^{s} N_i / \sum_{i=1}^{s} A_i$, where A refers to the area of the plaque, $\sum_{i=1}^{s} N_i$ refers to the total number of plaques and $\sum_{i=1}^{s} A_i$ refers to the total landscape area.

Landscape diversity index: $SHDI = -\sum_{i=1}^{s} P_i \ln P_i$

Landscape dominance: $D_h = H_{max} - SHDI$

$$= \ln S - \sum_{i=1}^{s} P_i \times \ln P_i$$

Landscape evenness: $SHEI = -\sum_{i=1}^{s} P_i \times \ln P_i / \ln S$

P_i = area ratio of type i landscape elements;
S = number of landscape types;
H_{max} = diversity index under the condition of maximum evenness.

The shape of landscape patches can reflect ecosystem stability and species dynamics. In general, the shape of landscape patches is the result of interference. To be exact, patches with a regular shape may suffer weak interference, while patches with an irregular shape may suffer from interference of different intensities and frequencies. Human deforestation activities increase fragmentation of the forest landscape, and patch shapes are directly related to human activities. Patches such as nature reserves have a relatively regular shape and better ecosystem stability due to weak interference. In contrast, for patches of the same area with a complex shape, the longer the perimeter, the more susceptible it is to outside interference. The shape is also of great importance to habitat selection and spread of species. For example, some birds can only survive in patches that simultaneously contain various habitats. The shape coefficient reflects the complexity of the patch shape. The larger the shape coefficient, the more complex the patch shape. Fractal dimension is one of the important aspects of fractal geometry applied in landscape ecology. Research has shown that fractal dimension exists widely in complex landscapes, generally with a value between 1 and 2. The larger the fractal dimension, the more complex the landscape.

Landscape connectivity has a great impact on ecosystem substance flow and energy flow. In this aspect, landscape elements with better connectivity, species spread, migration of individual organisms and energy flow are much easier and faster than landscape elements with worse connectivity. The connectivity index describes the connectivity among various patches of one landscape element. The larger the value, the better the connectivity among the patches. Landscape fragmentation describes the degree of fragmentation. The ecological effects would be very different between one large patch and various small patches formed due to fragmentation.

Over the entire landscape, the landscape diversity index describes the complexity of landscape composition; landscape dominance describes the relative advantages of various elements, and landscape evenness measures the evenness of landscape patches.

Evaluation of the green space system of the Luojiabian Interchange

Species composition, community structure and species diversity

The species composition of the Luojiabian Interchange green space system (excluding wetland plants) is shown in Table 9.1. There are a total of 20 varieties of species, including 11 varieties of woody plants, and 9 varieties of herbaceous plants, with the majority being lawn and ornamental flowers. In the woody plants, there are 9 varieties of broadleaf species and 2 varieties of coniferous species, with the most being broadleaf species. The majority of woody plants have a good self-sustaining capacity with no need for excessive artificial conservation, while ornamental flowers and lawns need require the investment of manpower and resources for conservation. There are 4 varieties of evergreen species and 16 varieties of deciduous species. All these species are

Table 9.1   Species and populations of the Luojiabian Interchange green space.

Species	Number of individuals	Growth type	Leaf nature	Necessity of artificial conservation
Cedrus deodara	6	Woody plant	Coniferous species	No
Metasequoia glyptostroboides	360	Woody plant	Coniferous species	No
Ligustrum lucidum	258	Woody plant	Broadleaf species	No
Diospyros kaki Thunb.	47	Woody plant	Broadleaf species	No
Liquidambar formosana	17	Woody plant	Broadleaf species	No
Sapium sebiferum	91	Woody plant	Broadleaf species	No
Pterocarya stenoptera	261	Woody plant	Broadleaf species	No
Salix babylonica	197	Woody plant	Broadleaf species	No
Nerium oleander	9414	Woody plant	Broadleaf species	Yes
Rhododendron pulchrum	7540	Woody plant	Broadleaf species	Yes
Sophora japonica	23	Woody plant	Broadleaf species	No
Canna indica	586 (m²)	Herbaceous plant		Yes
Cynodon dactylon	40 543 (m²)	Herbaceous plant		Yes
Poa praterzsis		Herbaceous plant		Yes
Lolium perenne		Herbaceous plant		Yes
Papaver rhoeas		Herbaceous plant		Yes
Echinacea purpurea		Herbaceous plant		Yes
Delphinium grandiflorum	6848 (m²)	Herbaceous plant		Yes
Cosmos bipinnatus		Herbaceous plant		Yes
Rudbeckia laciniata		Herbaceous plant		Yes

native species or alien species which can adapt to local habitats, proven by many years of transplantation. Overall, the species composition ratio is reasonable, which is conducive to exerting the ecological functions of the green space system. Evergreen species should be planted to ensure a large area of green space in autumn and winter, if necessary.

The Luojiabian Interchange green space has a total area of about $60\,614\,\mathrm{m^2}$, which accounts for a large proportion of the gross floor area of the interchange. However, the vast majority of existing plants are seedlings with a crown diameter not exceeding 3 m, a small leaf area index and low community canopy density. The total flux of substance flow and energy flow of the entire system is therefore still relatively small. With the development of a green space system, the ecological effects will be greatly enhanced. An integrated arrangement of trees, shrubs and herbaceous plants is considered during the construction of a green space. In addition, some other evergreen shrubs and native herbaceous plants which would not pose a competitive threat to the grass can be planted in the forest. The species diversity index of woody plants is 1.0133 (Shannon–Wiener index). According to research literature on the forest ecosystems of the Purple Mountain and Baohua Mountain, the species diversity index of woody plants in northern subtropical secondary forests is about 1.3. Since the species diversity of the Luojiabian Interchange green space is relatively low, the introduction of other woody plants which can meet the aforementioned criteria can be considered to increase the diversity.

Ecosystem functions

Productivity level: The Boreal Ecosystems Productivity Simulator (BEPS) model is employed to calculate the productivity of the Luojiabian Interchange green space system. The BEPS model is a process-based (physiology and ecology mechanism) distributed biogeochemical model developed by Professor Chen Jingming of the Canada Centre For Remote Sensing, which can be used to calculate regional NPP, evapotranspiration and other results using relatively simple entries (including meteorological data, leaf area index, land cover and soil moisture). The application of this model in Canada and East Asia (including China) has achieved satisfactory results. However, calculation with this model requires modifying certain main parameters according to the characteristics of native species to increase the applicability.

According to model calculation results, the average productivity level of coniferous trees for the Luojiabian Interchange green space system is $0.30\,\mathrm{kg}$ $\mathrm{C/m^2 \cdot a}$, and that of broadleaf trees is $0.52\,\mathrm{kg\,C/m^2 \cdot a}$. Such results are slightly smaller than the values in the summary of productivity observations on Chinese forest ecosystems by Academician Feng Zongwei. (The average productivity level of northern subtropical evergreen and deciduous broadleaf forests being $0.68\,\mathrm{kg\,C/m^2 \cdot a}$, and that of warm coniferous forests $0.42\,\mathrm{kg\,C/m^2 \cdot a}$.) The main reason is that most plants in the Luojiabian Interchange green space are seedling trees with few leaves, and their net assimilation of photosynthesis is smaller than that of adult standing trees. With the growth of seedling trees, the leaf area index increases and the NPP will increase accordingly. According to the observations made by researchers in Nanjing Purple Mountain and Jurong Baohua Mountain, the average leaf area index of a closed broadleaf forest is set as 5 in model computation; for coniferous forest it is 3.5 and 3 for grass. It can therefore be calculated that the annual carbon dioxide absorption capacity of the green space system at the Luojiabian Interchange is about 122.3 t, and the annual amount of oxygen released is 80.9 t.

Although not all of the carbon dioxide emissions from the combustion of fuel by motorway vehicles can be absorbed by the green space system, other harmful exhaust emissions are reduced on account of the adsorption function of the surrounding green space system. Currently, the productivity level of the Luojiabian Interchnage green space system is relatively low, and therefore auxiliary means such as fertilization and conservation are required to improve the productivity and promote substance circulation and energy flow of the entire ecosystem, thereby providing greater ecological value.

Species flow, substance circulation, energy flow, information flow and value flow: The dynamics of species in the artificial green space system depend on human activity. Species communities and populations can be constructed and maintained in accordance with the above principles. On the one hand, the conservation of functional species populations within the green space system should be promoted; on the other hand, the spread of invasive species should be contained.

The Luojiabian Interchange green space system is a relatively simple artificial ecosystem, and the process of species flow and energy flow is not complicated. Although the circulation of water and mineral elements serves as the basis for maintaining the entire system, in this particular green space system of the motorway it is of more practical significance to analyse the movement and transformation of pollutants in automobile exhaust emissions. The main issue to be considered is how to maximize the absorption and fixation of pollutants by organisms. Some people have used plants with high productivity and enriched with certain heavy metals for ecological restoration and have achieved remarkable results. The green space system of the motorway can take advantage of tailings ecological principles, and try to introduce species which can gather pollutants (for instance, *Sedum alfredii* can gather lead). Concentrated harvest and handling and the replanting of seedlings can be carried out after a certain period, which has an effect equivalent to reducing the pollutants discharged into the environment and the harm caused to humans.

Understanding information flow of green space systems can guide system construction and operation. To be specific, plants present different reactions to solar radiation. Some plant species have strong shade tolerance while other plant species are very tolerant to sunlight, meaning that it is important to rationally construct the layout of the system. Some coniferous species have an allelopathic effect on understory grass (meaning that the secreted chemical substances inhibit the growth of other plants), and therefore planting ornamental understory grass should be avoided. Although the main purpose of motorway green space systems is not economic, the ecological and economic benefits should be taken into consideration together in the current market economy in order to minimize the input–output ratio. The method of energy analysis can be used for simulating energy flow and value flow in the community so as to achieve the goal of scientific decision-making.

Decomposition: During the construction process of many artificial ecosystems, the role of decomposition has not been fully considered. For example, only the landscape aesthetic effect is considered in some green space systems, and measures such as clearing litter and moss layers are not conducive to the decomposition effect of microorganisms and soil animals, reducing community biodiversity, hindering the flow of ecosystem substances and energy as well as being adverse to exerting the service functions of ecosystems. During the maintenance of the Luojiabian Interchange green space, full attention should be paid to understory conservation and the improvement of system functions.

Landscape analysis

Although the absolute area of the Luojiabian Interchange green space is relatively small, it is of certain significance to carry out landscape ecological analysis on a small scale, and provide reference for overall planning and construction of the motorway. According to the main vegetation types of the green space system, the landscapes are classified into the following five landscape types: coniferous forest, broadleaf forest, grass, ornamental flowers and wetland plants. Landscape analysis software FRAGSTATS v3.3 was used to calculate the above landscape indices, with the results shown in Table 9.2.

Table 9.2   Landscape indices of the Luojiabian Interchange green space.

Landscape indices	Coniferous forest	Broadleaf forest	Grass	Flowers	Wetland plants
Shape coefficient	1.1135	1.1326	1.1789	1.1427	1.1618
Fractal index	1.0326	1.0671	1.1235	1.1114	1.1458
Connectivity index	0.2095	0.5581	0.4124	0.6347	0.2611
Landscape fragmentation	0.7564	1.1033	0.8869	0.8623	0.9556
Landscape diversity	1.1792				
Landscape dominance	1.9584				
Landscape evenness	0.8365				

The vegetation of interchange green spaces is planted artificially, with the shape of the patch being relatively regular, reflected by a low shape coefficient and fractal index in landscape measurement. The geometrical characteristics of the landscape elements presented by these two indices are basically identical: the shape of wetland plants and grass is relatively complex, followed by broadleaf forest and flowers, with coniferous forest ranking last. For patch connectivity, flowers and broadleaf forest rank first, followed by grass, and wetland plants and coniferous forest rank last; for landscape fragmentation, broadleaf forest ranks first, followed by wetland plants, flowers, grass and coniferous forest.

From the point of view of the entire landscape, the landscape diversity index of the green space system is 1.1792, belonging to a relatively simple landscape type, with a relatively high dominance of 1.9584, which is attributed to only a few types of landscape elements existing, with broadleaf trees and lawn being dominant in the landscape, as well as high landscape evenness, meaning that the overall landscape is of relative evenness. In summary, the Luojiabian Interchange green space is a relatively simple and typical artificial landscape, with both the landscape type and geometrical characteristics displaying low complexity. Such landscapes are strongly influenced by human activities, so scientific planning and rational construction of the green space is crucial for the health and development of the landscape.

Due to edge effects and other reasons, the ecological effects of one large patch and various small patches (with the total area equal to that of the large one) are different. In motorway green space systems, stretches of large green patches can greatly eliminate noise and pollutants, therefore it is necessary to increase connection patches among landscape elements with low connectivity (such as coniferous forest) as well as increase substance and energy exchange among these landscapes and improve system stability. In terms of the layout of the landscape, motorway vehicles are the source of carbon dioxide and other pollutants released into the atmosphere. According to the landscape security pattern theory proposed by Yu Kongjian, it is easy to find the strategic control area of such a green space system, which refers to both sides of the motorway. As reported in the relevant literature, the settlement range of lead in automobile exhaust fumes at both sides of the motorway is 50~100 m, so dense greenery is required within this range in order to control the lead settlement and reduce the impact on the surrounding farmland and residents. The Luojiabian Interchange green space system must still be strengthened in this regard.

The reconstruction of landscape patterns and analysis of landscape functions on a small scale can provide reference for overall motorway landscape plans on a large scale. The motorway, in terms of its large scale, can be regarded as a kind of landscape element of a corridor nature, with its main functions being the movement of human materials and the migration of human beings (transportation). When the green space on both sides of a motorway reaches a certain width, it can also serve as a corridor for the migration and spread of animal and plant species. For this reason, an increase in the width, coverage and connectivity of the green space is of great significance in terms of reducing pollutants and nature conservation.

Conclusions

According to ecological principles, investigation and analysis have been carried out on the current status of the Luojiabian Interchange green space. It has been found that the combination of species composition and community structure is reasonable; in future construction it is important to pay attention to increasing the amount of some evergreen broadleaf species, and further strengthen the synusia of the vertical structure of the community. Currently, the construction of the green space is still in its preliminary stages, and the overall productivity level is not high, meaning that the ecological effect cannot yet be brought into full play. A series of artificial conservation measures are required to accelerate the growth rate and increase the productivity, thereby enhancing the ecological functions of the entire system. Comprehensive consideration of all ecosystem functions such as species flow, substance circulation, energy flow and information value flow should be carried out to ensure all functions come into full play. At the landscape level, the connectivity of each patch in the green space needs to be increased; and in terms of the layout, greening at both sides of the highway needs to be strengthened to meet the requirements of the landscape security pattern. The Luojiabian Interchange green space system belongs to a simple artificial landscape with its development depending on human activity, meaning that scientific guidance in planning and construction is even more important.

9.2 Analysis of characteristics of green plant communities on typical sections

Objects for analysis

Five typical slope sections of the Nanjing-Hangzhou Motorway were selected, which are part of the landscape sensitive areas of this motorway. They have been carefully designed for the restoration of vegetation and landscape construction, and essentially reflect the design and greening of this motorway. In addition, the lengths of these five sections are basically the same, 960, 960, 975, 950 and 975 m, respectively. The area of the communities is also basically the same, with a surface area of 38 400, 38 400, 39 000, 38 000 and 39 000 m^2, respectively; this is conducive to comparison and meets the requirements of ecological research. These five typical sections are arranged from north to south on the Nanjing-Hangzhou Motorway, and possess clear regional characteristics.

Investigation and analysis methods

Species composition of the community

Any plant community is composed of certain plant species, so research on one plant community should first count the species composition of the community. In a plant community, each species has certain ecological amplitude, meaning a certain relationship with the environment and other species, and can reflect certain habitat characteristics as well as living conditions and evolution within the community. The research on a plant community should be carried out according to four plant types, namely trees, shrubs, ground cover plants and inter stratum plants.

Species quantity characteristics of the community

Quantitative analysis of species composition serves as the foundation of modern analytical methods for communities. Through an investigation of green vegetation at both sides of typical sections of the motorway and with reference to the original greening design drawings, the following information of the sample community is recorded: diameter at breast height, crown diameter and height of trees as well as the number of each species; and height and crown diameter of shrubs as well as the number of each species. The basic characteristic parameters of the community are therefore obtained, with the main features given in the following.

Abundance: This refers to an individual number of a certain species in a sample plot, which is an indicator of the number of individuals of different species.

Density: This refers to an individual number of unit area or unit space. Trees and shrubs are counted by plant.

$$D = \frac{N}{S} \tag{9.1}$$

$D =$ density;
$N =$ individual number of certain plant in sample plot;
$S =$ area of sample plot, calculated by $100\,m^2$.

Density ratio: This refers to the ratio of the density of a certain species in the sample plot to the density of the species in the sample plot with the highest density.

Prominence: This is calculated as the ratio of the sectional area at breast height (1.3 m from the ground) of a certain tree species to all of the sectional area within the sample plot.

Prominence ratio: This refers to the ratio of the prominence of a certain species in the sample plot to the greatest prominence among all the species in the sample plot.

Summed dominance ratio: This is a comprehensive quantitative indicator representing the status and role of a certain species in the community, involving density ratio, prominence ratio, frequency ratio, height ratio and weight ratio, with four categories of two factors, three factors, four factors and five factors.

The following methods are adopted herein:

$$\text{Trees } SDR2 = \frac{\text{Density ratio} + \text{Prominence ratio}}{2} \times 100\% \qquad (9.2)$$

$$\text{Shrubs } SDR2 = \frac{\text{Density ratio} + \text{Height ratio}}{2} \times 100\% \qquad (9.3)$$

Community life-form

Life-form refers to the type reflected in appearance after long-term adaptation to consolidated environmental conditions. The life-form system, proposed by the famous Danish ecologist C. Raunkiaer, selects the position of dormant buds during seasons with adverse conditions as the standard for the classification of life-forms. Higher plants can be divided into the following five life-forms:

(1)   Phanerophytes: the dormant bud is more than 25 cm above the ground; and can be divided into four sub-types according to the height.
(2)   Chamaephytes: the renewal bud is located above the soil surface and less than 25 cm from the ground; these are mostly shrubs, subshrubs or herbs.
(3)   Hemicryptophytes: the renewal bud is in the soil layer near the ground, and the part above the ground withers in winter; these are mostly perennial herbs.
(4)   Cryptophytes: the renewal bud is in a deep soil layer or water; these are mostly bulb, tuber, rhizomes and perennial herbs or aquatic plants.
(5)   Therophytes: these plants live through the winter as seeds.

The percentage of each life form species in a community makes up the life-form spectrum of the community. Each plant community consists of plants of several life-forms, with one dominant life-form. In general, when phanerophytes are dominant, it reflects hot and humid characteristics during the plant-growing season in the area where the community is located; when hemicryptophytes are dominant, it reflects a long cold season; when geophytes are dominant it reflects a cold and wet environment; and when therophytes are most abundant, it indicates an arid climate.

The life-form spectrum of five typical sections obtained by applying the life-form system of C.Raunkiaer is used herein for analysis.

Percentage of a certain life-form =

$$\frac{\text{Number of plant species of a certain life-form in the community}}{\text{Number of all plant species in the community}} \times 100\%$$

(9.4)

Community species diversity

Species diversity of a plant community refers to the number of species in a community and the distribution evenness of each species. It not only reflects the species richness of a community composition but also reflects the relationship between different natural and geographical conditions and the community, as well as the stability of the community, making it an important feature of the community structure. The main indicators are given in the following.

Species richness index (S): Species richness refers to the number of species, and is the simplest and most ancient method used for measuring species diversity.

Simpson index:

$$D = 1 - \sum P_i^2$$

where

$$P_i = \frac{N_i}{N}. \tag{9.5}$$

Shannon–Wiener Index:

$$H' = - \sum P_i ln P_i$$

where

$$P_i = \frac{N_i}{N}. \tag{9.6}$$

Pielou evenness index:

$$\text{Based on the Simpson index}: J_{si} = \frac{1 - \sum P_i^2}{1 - 1/S} \tag{9.7}$$

$$\text{Based on the Simpson–Wiener index}: J_{si} = \frac{- \sum P_i \ln P_i^2}{1nS} \tag{9.8}$$

N_i = the individual number of the ith species in the sample plot;
N = the individual number of all plants in the sample plot.

Community similarity comparison

The simplest mathematical representation of plant community similarity is the community coefficient proposed by Jaccard, which is expressed by the ratio of common species to all species in two plant communities. Species similarity refers to the similarity of plant species composition among communities or sample plots, and it is an important foundation for community analysis.

$$IS_j = \frac{c}{a + b + c} \times 100\% \qquad (9.9)$$

C = the number of common species;
A = the recorded number of unique species in the first sample plot;
B = the recorded number of unique species in the second sample plot.

Results and analysis

Analysis of species composition and number of green communities at five typical sections

The quantitative characteristics of green plant community species at typical section 1 (K2+600~K3+560), which is near the Luojiabian Interchange on the Nanjing-Hangzhou Motorway, are considered.

As seen from Table 9.3, the tree layer at typical section 1 has a total of 10 varieties of plants, belonging to 9 families and 10 genera, including the three evergreen plants *Ligustrum lucidum*, *Cinnamomum camphora* and *Pinus elliottii*; the rest are deciduous trees.

Table 9.3   Species quantitative characteristics of tree layer at typical section 1.

Name	Abundancy (plants)	Density (plant/ 100 m^2)	Density ratio (%)	Prominence ratio (%)	Summed dominance ratio (%)	Order of summed dominance ratio
Ligustrum lucidum	123	0.32	0.98	1	99.2	1
Zelkova schneideriana	125	0.33	1	0.89	94.66	2
Cinnamomum camphora	94	0.24	0.75	0.97	85.96	3
Sapium sebiferum	37	0.1	0.3	0.3	29.84	4
Acer buergerianum	23	0.06	0.18	0.19	18.55	5
Koelreuteria integrifoliola	17	0.04	0.14	0.17	15.55	6
Taxodium ascendens	15	0.04	0.12	0.08	10.03	7
Pinus elliottii	11	0.03	0.09	0.05	6.915	8
Salix babylonica	6	0.02	0.05	0.05	4.84	9
Metasequoia glyptostroboides	3	0.01	0.02	0.02	2.42	10
Total	454	1.18				

The tree species at this typical section are at a height of 2.5~5 m; the layers are not significant and diameter at breast height is less than 10 cm. In general, the tree size is small, the polarization in individual numbers is very obvious, and *Ligustrum lucidum*, *Zelkova schneideriana* and *Cinnamomum camphora* account for the majority, amounting to 123, 125 and 94 trees, respectively. The density of *Zelkova schneideriana* ranks first, at 0.33 plant/100 m². The numbers of the rest of the tree species are less than 40, with *Salix babylonica* and *Metasequoia glyptostroboides* ranking last, at 6 and 3, respectively.

The summed dominance ratio reflects the dominance of each species in the community, and its ecological significance and importance values are equivalent. The whole section is regarded as a community in this investigation, with no quadrat set, meaning that there is no frequency, so the summed dominance ratio is taken as the indicator for measuring species dominance in the community. As seen from the summed dominance ratio, the tree layer at typical section 1 has two dominant species, *Ligustrum lucidum* and *Zelkova schneideriana*, and *Cinnamomum camphora* is the second most dominant species.

As seen from Table 9.4, the shrub layer at typical section 1 has a total of 10 varieties of plants, belonging to 9 families and 9 genera, including the five evergreen plants *Jasminum mesnyi*, *Nerium oleander*, *Pittosporum tobira*, *Gardenia jasminoides* and *Aucuba chinensis*, and one semi evergreen plant, *Rhododendron pulchrum*, with the remaining four being deciduous plants. The shrub layer at typical section 1 is divided into two layers according the height, with *Nerium oleander*, *Hibiscus syriacus*, *Jasminum mesnyi*, *Gardenia jasminoides* and *Pittosporum tobira* forming the first layer at an average height of 70–110 cm, and the rest forming the second layer at a height of less than 50 cm. Looking at the horizontal structure, the density of *Rhododendron*

Table 9.4 Species quantitative characteristics of shrub layer at typical section 1.

Name	Abundance (plants)	Density (plant/ 100 m²)	Density ratio (%)	Prominence ratio (%)	Summed dominance ratio (%)	Order of summed dominance ratio
Rhododendron simsii	5259	13.7	1	0.32	65.9	1
Nerium oleander	1546	4.03	0.29	1	64.7	2
Rhododendron pulchrum	4495	11.71	0.44	0.32	58.6	3
Hibiscus syriacus	2288	5.96	0.43	0.73	58.1	4
Jasminum mesnyi	1208	3.15	0.23	0.82	52.4	5
Pittosporum tobira	1114	2.9	0.21	0.64	42.4	6
Gardenia jasminoides	743	1.93	0.14	0.68	41.2	7
Bacca	1369	3.57	0.26	0.5	38	8
Hypericum monogynum	1065	2.77	0.2	0.41	30.6	9
Aucuba chinensis	214	0.56	0.04	0.5	27	10
Total	19 301	50.3				

simsii and *Rhododendron pulchrum* in the Ericaceae family is the greatest, at more than 10 shrubs/100 m^2; the density of *Aucuba chinensis* is the lowest, at just 0.56 shrubs/100 m^2.

Although the individual number of *Rhododendron pulchrum* is second only to *Rhododendron simsii*, being the second most abundant plant in the shrub layer at typical section 1, the dominant species of this shrub layer are *Rhododendron simsii* and *Nerium oleander*, which is mainly because *Nerium oleander* is in the first layer of this shrub layer.

After comprehensive analysis, there are 30 varieties of species at typical section 1, including 20 varieties of woody plants and 10 varieties of herbaceous plants. Among the woody plants, there are 18 varieties of broadleaf species, 2 varieties of coniferous species, 8 varieties of evergreen plants, 1 kind of semi evergreen plant and 11 varieties of deciduous plants.

Life-form spectrum analysis of communities at five typical sections

According to Table 9.5 and Figure 9.1, the proportion of phanerophytes at the five typical sections is less than 70%, and there is a shortage of tall and medium phanerophytes. There are relatively few chamaephytes at the five typical sections. Among the five typical sections, the life-form spectrum of typical sections 1, 2 and 4 are similar, and that of typical sections 3 and 4 are similar. Comparing the Raunkiaer life-forms at adjacent areas (subtropical evergreen broadleaf forest at Mount Huangshan and in Zhejiang Province, and temperate evergreen broadleaf forest on the north slopes of the Qinling Mountains), the Raunkiaer life-forms of the subtropical evergreen broadleaf forest at Mount Huangshan and Zhejiang Province are as follows: phanerophytes>hemicryptophytes>cryptophytes>therophytes>chamaephytes; and the Raunkiaer life-forms of the temperate evergreen broad-leaf forest on the north slopes of the Qinling Mountains are as follows: phanerophytes>hemicryptophytes>chamaephytes>cryptophytes>therophytes. From north to south, the proportions of phanerophytes, cryptophytes and therophytes increase, with phanerophytes being the dominant species, while the proportions of chamaephytes and hemicryptophytes decrease. The Nanjing-Hangzhou Motorway is located at northern subtropical and extends from north to south, and the five typical sections are arranged from north to south. The Raunkiaer life-forms of the motorway should therefore be identical to the changes from the temperate evergreen broadleaf forest on the north slopes of the Qinling Mountains to the subtropical evergreen broadleaf forest in Zhejiang Province, while the life-form spectrum of the five typical sections should be: phanerophytes>hemicryptophytes>therophytes>cryptophytes>chamaephytes. Upon comparison, it was found that the life-form of the typical sections does not reflect such a trend, and that the proportion of phanerophytes at typical sections 3 and 4 is smaller than that at typical sections 1 and 2, the proportion of hemicryptophytes is increased, and no chamaephytes exist at any of the five typical sections. The life-form structure

Table 9.5 Life-form spectrum comparison of communities at five typical sections and other areas.

Life-form	Typical section 1/TS 1	Typical section 2/TS 2	Typical section 3/TS 3	Typical section 4/TS 4	Typical section 5/TS 5	Temperate deciduous broadleaf forest on the north slopes of the Qinling Mountains	Mount Huangshan	Subtropical evergreen broadleaf forest in Zhejiang Province
MaPh. (>30 m)	0/20	0/21	0/22	0/22	0/23	–	–	–
MePh. (8–30 m)	0/20	0/21	0/22	0/22	0/23	–	–	–
MiPh. (2–8 m)	10/20	10/21	11/22	11/22	12/23	–	–	–
NPh. (0.25–2 m)	10/20	11/21	11/22	11/22	12/23	–	–	–
Ph. (%)	66.67	67.74	57.90	59.46	69.70	52	72.5	76.1
Ch. (%)	0.07	0.06	0.05	0.05	0.06	5.0	1.5	1.0
H. (%)	13.33	12.9	26.32	24.32	12.12	38	18.1	13.1
Cr. (%)	3.33	3.23	2.63	2.70	3.03	3.7	5.9	7.8
T. (%)	10.00	9.68	7.89	8.11	9.09	1.3	2.0	2.0
Total	30	31	38	37	33			

MaPh., megaphanerophytes; MePh., mesophanerophytes; MiPh., microphanerophytes; NPh., nanophanerophytes; Ph., phanerophytes; Ch., chamaephytes; H., hemicryptophytes; Cr., cryptophytes; T., therophytes.

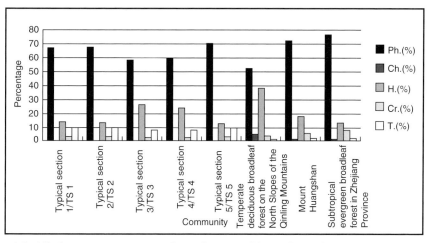

Figure 9.1   Life-form spectrum comparison of communities at five typical sections and other areas.

at the five typical sections is therefore different from natural vegetation under a stable state. The conservation costs will inevitably rise for maintaining the existing life-form structure.

Diversity analysis of species at five typical sections

Based on the above calculation method, Microsoft Excel was used for processing raw data in order to obtain the species diversity index of vegetation communities at the five typical sections (Tables 9.6 and 9.7).

Table 9.6   Species diversity indices comparison of tree layers at the five typical sections.

Position	Species richness (S)	Simpson index	Shannon–Wiener index	Pielou index (based on Simpson index)	Pielou index (based on Shannon–Wiener index)
Typical section 1/ TS 1	10	0.795	1.807	0.884	0.785
Typical section 2/ TS 2	9	0.851	2.012	0.957	0.916
Typical section 3/ TS 3	9	0.855	2.050	0.962	0.929
Typical section 4/ TS 4	9	0.857	2.101	0.964	0.933
Typical section 5/ TS 5	12	0.853	2.13±0.07	0.930	0.956
Mid-subtropical evergreen broadleaf forest (He *et al.*, 1998)					
Evergreen broadleaf forest in Zhejiang Province (Yu, 2003)	15.05±7.01	0.7887			

Table 9.7 Species diversity indices comparison of shrub layers at the five typical sections.

Position	Species richness (S)	Simpson index	Shannon–Wiener index	Pielou index (based on Simpson index)	Pielou index (based on Shannon–Wiener index)
Typical section 1/TS 1	10	0.834	2.010	0.927	0.873
Typical section 2/TS 2	11	0.857	2.143	0.943	0.894
Typical section 3/TS 3	7	0.750	1.579	0.875	0.811
Typical section 4/TS 4	7	0.592	1.263	0.691	0.649
Typical section 5/TS 5	10	0.807	1.828	0.896	0.794
Evergreen broadleaf forest in Zhejiang Province (Yu, 2003)	28.83±20.69				

Tree layer: Compared with the Shannon–Wiener index 2.13 ± 0.07 of mid-subtropical evergreen broadleaf forest (He *et al.*, 1998), the species diversity of tree layers at the five typical sections is slightly lower, while the Simpson index of the five typical sections is significantly higher than the Simpson index of the tree layer of evergreen broadleaf forest in Zhejiang Province, according to species diversity research by Yu Shuquan in 2003 (Yu, 2003). Considering the area of the quadrat investigated by Yu Shuquan was only $400 \, \text{m}^2$, the number of species within the quadrat was much less than that within each typical section, therefore it is not surprising that the Simpson index is less than that of the five typical sections. After all, the Simpson index considers the number of species in the community and the diversity index of each relative abundant species. However, based on a comparison between the species richness index of five typical sections and the average value of the tree layer of evergreen broadleaf forest in Zhejiang Province, it can be seen that the number of species in the tree layers at the five typical sections is relatively low.

Based on the above analysis, it can be seen that the variety of species in the tree layers at the five typical sections is not abundant but the number of individual species is large. The diversity of tree species is not high overall.

Shrub layer: According to the species diversity of shrub layers of the five typical sections in Table 9.7, the species richness index presents two types of distribution: the index of typical sections 1, 2 and 5 is greater than 10, amounting to 10, 11 and 10, respectively; that of typical sections 3 and 4 is less than 10, at 7. However, it is significantly lower when compared with that of shrub layers of evergreen broadleaf forest in Zhejiang Province. The Simpson index and Shannon–Wiener index of shrub layers show high consistency, with the diversity in descending order as: typical section 2>typical section 1>typical section 5>typical section 3>typical section 4. The descending order of the evenness index J_{si}, which is based on the Simpson index, is as follows: typical section 2>typical section 1>typical section 5>typical section 3>typical section 4; while the descending order of the evenness index J_{sw}, based on the Shannon–Wiener index, is as follows: typical section 2>typical section 1>typical section 3>typical section 5>typical section 4.

General characteristics of vegetation communities at five typical sections

The total number of species of each community at the five typical sections is more than 30. To be exact, typical section 3 has the most varieties of species, with 38 varieties of plants; typical section 4 ranks second which has a similar plant composition to typical section 3 and has 37 varieties of plants. There are 6 varieties of lianas at these two sections. Typical section 1 has the least number of species, at 30 in total, with no liana. Typical sections 2 and 5 have 32 and 33 varieties of species; each has one liana. The woody species at the five typical sections amount to more than 20 varieties. To be exact, typical section 5 has the most abundant woody plant species, with 23 varieties; typical section 1 has the least woody plant species, with 20 varieties; typical sections 2, 3 and 4 have woody plant species of 21, 22 and 22 varieties, respectively. Typical section 3 has the most abundant herbaceous plant species, with 16 varieties; typical section 4 ranks second, with 15 varieties. Typical section 5 has the most abundant woody plant species and the least herbaceous plant species, with 10 varieties. Typical section 1 has the same number of herbaceous plant species as typical section 5.

As seen from Table 9.8, which is a species comparison of trees, shrubs and herbs at the five typical sections, the species ratio of trees, shrubs and herbs at typical sections 1, 2 and 5 is close to 1:1:1, and that of typical sections 3 and 4 is less than 1:1:1. In addition, the number of herb species is larger than the number of trees and shrubs.

According to tree layer analysis of each community, the number of evergreen tree species of the five communities is lower than that of deciduous

Table 9.8 **Species composition comparison of the five typical sections.**

Item		Unit	Typical section 1	Typical section 2	Typical section 3	Typical section 4	Typical section 5
Woody plant:herbaceous plant		Variety	20:10	21:11	22:16	22:15	23:10
Tree:shrub:herb		Variety	10:10:10	9:11:10	9:7:16	9:7:15	12:10:10
Liana		Variety	–	1	6	6	1
Tree	Evergreen:deciduous	Variety	3:7	2:7	3:6	2:74:8	
	Evergreen:deciduous		228:/226	248:471	114:441	77:485	617:692
	Broadleaf:coniferous	Variety	7:3	8:1	7:2	7:2	10:2
	Broadleaf:coniferous		425:29	682:37	451:104	490:72	1169:140

tree species. For evergreen tree species, typical section 5 has the largest number with 4 varieties, typical sections 1 and 3 have 3 varieties each, and typical sections 2 and 4 have 2 varieties each. For deciduous tree species, typical section 5 has the largest number with 8 varieties, typical sections 1, 2 and 4 have 7 varieties each, and typical section 3 has the least number with 6 varieties. In terms of the quantity of plants, the quantity ratio and the number of species for typical sections 2, 3 and 4 show a positive correlation, and the quantity of evergreen trees is less than that of deciduous trees. The quantity ratio of evergreen trees and deciduous trees at typical sections 1 and 5 is close to 1:1, and the quantity of evergreen trees at typical section 1 is slightly larger than that of deciduous trees. In addition, the quantity and the number of species of broadleaf trees are larger than those of coniferous trees. For the number of coniferous tree species, typical section 1 has the largest number with 3 varieties, while typical section 2 has the least number with 1 variety; the remaining three typical sections have 2 varieties each. For the quantity of coniferous tree plants, typical section 5 has the largest quantity with 140 plants, while typical section 1 has the least quantity with only 29 plants.

As we can be seen from Table 9.9, the total density of trees at the five typical sections is less than 10 plants/100 m^2. To be specific, typical section 5 has the highest density with 3.42 plants/100 m^2, and that of the rest of the four typical sections is 1–2 plants/100 m^2, with typical section 1 ranking last at 1.18 plants/100 m^2. Regarding the density of shrubs, typical section 5 has the highest density, with 81.62 plants/100 m^2; typical section 1 ranks second with 50.3 plants/100 m^2, while typical sections 2, 3 and 4 have a density less of than 25 plants/100 m^2. It can therefore be seen from the above comparison that typical section 5 has the highest density of plants, while typical sections 3 and 4 have the lowest density, but overall, the tree density of the five typical sections is relatively low. The current characteristics shown in Table 9.9 represent community types of savanna and open grassland occupying a large area. The zonal vegetation in Jiangsu Province consists of evergreen and deciduous mixed forests or deciduous broadleaf forests with evergreens. Given the climatic conditions in Jiangsu Province, there are bound to be other invasive species. Upon invasion, settlement, reproduction and evolution of species, the existing communities would be gradually converted into zonal vegetation of this region. In order to maintain the current green design style, the invasive trees and weeds should be cleared away, inevitably increasing conservation costs.

Table 9.9 **Species structural characteristics comparison of the five typical sections.**

Item	Unit	Typical section 1	Typical section 2	Typical section 3	Typical section 4	Typical section 5
Total density of trees	Plant/100 m^2	1.18	1.87	1.45	1.47	3.42
Total density of shrubs	Plant/100 m^2	50.3	24.2	24.06	24.13	81.62
Indigenous:garden	Kind	12:20	17:14	13:25	13:24	14:19

For the ratio of indigenous plants and garden plants at the five typical sections, it can be seen from Table 9.9 that the proportion of indigenous plants at typical section 2 is higher than that of garden plants, while the introduced garden plants of the remaining sections account for a large proportion, which is much higher than that of indigenous plants. From the above statistics, it can be seen that the plant design of the five typical sections is more similar to an urban landscape green design, meaning that more garden plants are used. Such plant arrangement does not seem natural in the surrounding environment, and fails to integrate naturally into the environment. After comprehensive analysis, the types of plant communities in the five typical sections are similar, with no dense forest, deciduous evergreen mixed forest, evergreen broadleaf forest or shrubs. In addition, landscape heterogeneity is low.

Community similarity comparison

The following are common species of plant communities at the five typical sections: *Zelkova schneideriana, Cinnamomum camphora, Nerium olean-der, Bacca Hypericum monogynum, Jasminum mesnyi, Cynodon dactylon, Orychophraqmus violaceus, Lolium perenne, Viola odorata, Herba taraxaci, Chrysanthemum indicum, Mirabilis jalapa* Linn., *Coreopsis lanceolata* and *Oenothera erythrosepala* Borb.

As we can be seen from Table 9.10 and Table 9.11, the similarity coefficient of community species of typical sections 3 and 4 is the largest (92.3%), followed by typical sections 2 and 5 (77.8%), while that of typical sections 4 and 5 is the smallest (32.1%). In addition, the similarity coefficient between the five typical sections cannot present the following characteristics well according to their geographical distribution: the longer the distance between typical sections, the larger the similarity coefficient; the shorter the distance between typical sections, the smaller the similarity coefficient. Furthermore, the similarity coefficient between the five typical sections is larger than the coefficient of 22.1% between Nanjing and Yixing (Zhu and Feng, 1998), which indicates that species similarity among plant communities of the five typical sections is relatively large.

Table 9.10   Number of common species of the five typical sections.

Position	Typical section 1	Typical section 2	Typical section 3	Typical section 4	Typical section 5
Typical section 1	–	25	17	17	27
Typical section 2		–	8	18	28
Typical section 3			–	36	18
Typical section 4				–	17
Typical section 5					–

Table 9.11   Community similarity index of the five typical sections.

Position	Typical section 1	Typical section 2	Typical section 3	Typical section 4	Typical section 5	Nanjing
Typical section 1	–	69.5	38.6	34	75.0	
Typical section 2		–		35.3	36	77.8
Typical section 3			–		92.3	34
Typical section 4				–		32.1
Typical section 5					–	
Yixing						22.1

Conclusion

By applying the theories and methods of vegetation ecology, based on analysis and comparison of plant communities' characteristics on the five typical sections, using species composition and quantitative characteristics of the species as the main evaluation indicators, the structural characteristics of tree layers and shrubs layers of the plant communities of the five typical sections can be clearly described, and the dominant species of tree layers and shrub layers within each community as well as their vertical structure and horizontal structure can be determined.

According to above species composition analysis of the five typical sections, broadleaf species constitute the vast majority of woody plants at the five typical sections, and are in general native species or alien species which can adapt to the local habitat, proven by many years of transplantation. However, the species in the vegetation composition at the five typical sections are similar, with the species composition of typical sections 1, 2 and 5 being relatively similar, and those of typical sections 3 and 4 being relatively similar. In general, the species composition varies little among the five typical sections.

The status of each species in the community is analysed according to the order of the summed dominance ratio. The following species are the dominant species of each typical section: *Ligustrum lucidum* and *Zelkova schneideriana* for typical section 1, *Zelkova schneideriana* for typical section 2, *Sapium sebiferum* for typical section 3, *Sapium sebiferum* for typical section 4, and *Ligustrum lucidum* for typical section 5.

In terms of the arrangement of species composition, typical sections 1, 2 and 5 are relatively similar, while typical sections 3 and 4 are relatively consistent, which is partially due to the similarity of species composition among the five typical sections. It also indicates that the interspecific relationship between tree layers on the five typical sections is relatively similar and lacks change.

In terms of the tree layers and shrub layers of the five typical sections, the species diversity is not significant, the species are not rich and the composition is relatively similar. For tree layers, typical section 1 has 10 varieties of plants belonging to 9 families, typical section 2 has 9 varieties of plants belonging to 9 families, typical section 3 has 9 varieties of plants belonging to 8 families, typical section 4 has 9 varieties of plants belonging to 8 families, and typical section 5 has 12 varieties of plants belonging to 12 families. There are 4 common species among the five typical sections, which belong to 3 families. For shrub layers, typical section 1 has 10 varieties of plants belonging to 9 families, typical section 2 has 11 varieties of plants belonging to 11 families, typical section 3 has 7 varieties of plants belonging to 5 families, typical section 4 has 7 varieties of plants belonging to 5 families, and typical section 5 has 10 varieties of plants belonging to 9 families. There are 3 common species among the five typical sections, which belong to 3 families. In addition, the Shannon–Wiener index of the tree layers is generally lower than that of the natural vegetation in the surrounding areas.

The analysis on species composition and quantitative characteristic indices of species describe the vegetation structure of each community (of typical sections). As longitudinal analysis, these indices can only show the status of each community. The Raunkiaer community life-form system, species diversity index and community similarity can therefore be used to horizontally analyse the structure of plant communities of the five typical sections. The relevant indices of the adjacent areas can be used for comparison. After analysis, it has been found that the Raunkiaer community life-form of the five typical sections is very similar, displaying no expected changing trend from north to south. Furthermore, compared with the Raunkiaer community life-form of the adjacent areas, it has been found that the life-form of the five typical sections is slightly different, meaning that conservation costs will inevitably need to increase in order to maintain the current life-form.

After analysis of the five typical sections using the species diversity index S, Simpson index, the Shannon–Wiener index and Pielou evenness indices J_{si} and J_{sw}, it has been shown that the species diversity of tree layers and shrub layers of each community is relatively low; there is also relatively good consistency between the indices.

After community similarity comparison, it has been found that typical sections 1, 2 and 5 have a relatively similar species composition, while typical sections 3 and 4 have a relatively similar species composition. The minimum similarity coefficient between the five typical sections (32.1%) is also larger than that (22.1%) between Nanjing and Yixing, indicating that plants selected for the arrangement of each typical section are relatively similar. After comprehensive analysis, the arrangement of plant species of the five typical sections does not display an expected changing trend from north to south.

Through the above longitudinal and horizontal comparisons, the appearance of plant communities of the five typical sections can be clearly outlined. However, the result is not satisfactory. Whilst the species selected for the Nanjing-Hangzhou Motorway can generally adapt to site conditions of the area along the motorway and grow normally, there are some common defects amongst the plants communities of the five typical sections. These include low species diversity and lack of an appropriate community structure, with few tree species (especially evergreen broadleaf tree species and coniferous tree specious) and many semi-shrubs or herbs. In addition, the species composition of the five typical sections is extremely similar, resulting in a relatively unvaried appearance of the communities. It is recommended to improve the vegetation structure of these sections and pay attention to these issues during future landscape design in order to improve the quality of the landscape and design level of the motorway.

9.3   Green space system of slopes of typical sections

Motorways are huge man-made constructions. Their construction and operation upon completion have a negative impact on the surrounding environment. It is therefore crucial to handle the relationship between construction and environmental protection well. The company URS/Scott Wilson has carried out landscape design on the slopes of some typical sections along the Nanjing-Hangzhou Motorway. Various methods have been adopted to integrate the motorway with the surrounding environment, such as framed planting for creating a large area of natural landscape, hedgerow planting for achieving a neat, symmetrical, natural and harmonious effect at some linear areas or narrow sections, and barrier-type planting, which uses high-density planting to mitigate the impact of motorway noise on residents. These designs result in new ideas and methods for the creation of a motorway landscape, which can improve the aesthetic appearance of the motorway and also provide an alternative visual experience to travellers. However, ecological analysis should be carried out to determine whether this design is scientific and reasonable in terms of plant selection, plant combination and plant community structure.

The theories and methods of vegetation ecology are adopted in this section for implementing preliminary analysis and study of the slope green space system of the typical sections.

Main evaluation content and methods

The five typical sections in the engineering design for Phase I of the Nanjing-Hangzhou Motorway are taken as examples. Each section is regarded as a separate plant community, which can also be called a small artificial ecosystem. An investigation of the vegetation should be carried out and combined with the original green design drawings. Tree and shrub layers in a sample community are selected for recording raw data. The basic characteristic parameters are obtained according to raw data, including species composition, density, coverage and abundance. On this basis, further analysis should be implemented in terms of species diversity and life-forms of the community. These communities are then compared with similar communities of natural vegetation in the surrounding environment.

Ecological evaluation of vegetation in the typical sections

Quantitative and characteristic analysis of the population

(1) Taking typical section 1 as an example, statistical indices such as the species quantity are listed in Tables 9.12 and 9.13).

(2) Quantitative characteristics of community species composition are listed from Tables 9.12–9.16.

Table 9.12   The composition of trees in the five typical sections.

Name	Trunk diameter, D (cm)	Height, H (cm)	Crown diameter, S (cm)	Quantity (plant)
Cinnamomum camphora	9	425	300	94
Ligustrum lucidum	8	325	200	123
Pinus elliottii	6	250	200	11
Acer buergerianum	5	400	200	23
Koelreuteria integrifoliola	465	275	17	
Sapium sebiferum	8	300	200	37
Metasequoia glyptostroboides	350	150	3	
Taxodium ascendens	6.5	300	100	15
Zelkova schneideriana	7.5	300	250	125
Salix babylonica	8	250	200	6
Total				454

Table 9.13   The composition of shrubs in the five typical sections.

Name	Height, H (cm)	Crown diameter, S (cm)	Quantity (plant)
Aucuba chinensis	55	35	214
Gardenia jasminoides	75	55	743
Hibiscus syriacus	80	60	2288
Hypericum monogynum	45	35	1065
Jasminum mesnyi	90	90	1208
Kerria japonica	55	35	1369
Nerium oleander	110	55	1546
Pittosporum tobira	70	70	1114
Rhododendron pulchrum	35	35	4495
Rhododendron simsii	35	25	5259
Total			19 301

Table 9.14   The composition of ground cover in the five typical sections.

Shrub types	Name	Seeding quantity (g/100 m²)	Percentage (%)	Area (m²)
Turf mixed seeding	Cynodon dactylon	1000	70	
	Poa	400	20	8300
	Lolium perenne	550	10	
	Orychophraqmus violaceus	30	25	
	Viola odorata	3	15	
	Herba taraxaci	30	10	
Wildflower mixed seeding	Chrysanthemum indicum	50	10	3224
	Mirabilis jalapa Linn.	50	20	
	Coreopsis lanceolata	5	10	
	Oenothera erythrosepala Borb.	2	10	
	Total			11 524

Table 9.15   The features of trees in the five typical sections.

Name	Density (plant/ 100 m²)	Density ratio	Prominence	Prominence ratio	Height ratio	Summed dominance ratio, SDR3 (%)
Cinnamomum camphora	0.752	0.19	0.97	0.914	87.773	
Ligustrum lucidum	0.392	0.984	0.197	1	0.699	89.431
Pinus elliottii	0.035	0.088	0.01	0.05	0.538	22.531
Acer buergerianum	0.073	0.184	0.037	0.19	0.86	41.04
Koelreuteria integrifoliola	0.136	0.034	0.17	1	43.697	
Sapium sebiferum	0.118	0.296	0.059	0.3	0.645	41.399
Metasequoia glyptostroboides	0.01	0.024	0.005	0.02	0.753	26.703
Taxodium ascendens	0.048	0.12	0.016	0.08	0.645	28.189
Zelkova schneideriana	0.398	1	0.176	0.89	0.645	84.612
Salix babylonica	0.019	0.048	0.01	0.05	0.538	21.147
Total	1.446		0.734	3.73		

Diversity analysis of community species

Species composition: Population abundance and density serve as the main indicators of community composition. The biological characteristics of the population should be used to determine a reasonable planting density in order to avoid vicious competition among species due to high density, which can result in plant death. Multiple populations within a community can overlap and compensate each other due to their different ecological niches. The community can only be stable with a reasonable structure.

Table 9.16   The features of shrubs in the five typical sections.

Name	Density (plant/ 100 m²)	Density ratio	Height ratio	Summed dominance ratio, SDR2 (%)
Aucuba chinensis	0.682	0.041	0.5	27
Gardenia jasminoides	2.366	0.141	0.682	41.2
Hibiscus syriacus	7.287	0.435	0.727	58.1
Hypericum monogynum	3.392	0.203	0.409	30.6
Jasminum mesnyi	3.847	0.23	0.818	52.4
Kerria japonica	4.36	0.26	0.5	38
Nerium oleander	4.294	0.294	1	64.7
Pittosporum tobira	3.548	0.212	0.636	42.4
Rhododendron pulchrum	14.32	0.855	0.318	58.6
Rhododendron simsii	16.75	10.318	65.9	
Total				

Tables 9.12–9.16 show the species composition of the slope green space of the five typical sections. The number of species at each section is generally approximately 30, with the vast majority being broadleaf species. The proportion of evergreen species and deciduous species is similar to local northern subtropical natural vegetation; they are mainly native species or exotic species that have adapted to the local habitat. However, there are relatively few coniferous trees, and there are no coniferous trees at typical section 2. The problem lies in the evergreen woody plants, with shrubs significantly more than trees. The tree layer of the community is sparse and monotonous, and the morphological structure is not perfect. In addition, most of the woody plants are cultivated flowers and grass and there is a lack of natural-style grass. In cases such as these, more human and material resources are required in order to maintain such features, otherwise natural successional changes will occur.

Population diversity: The Shannon–Wiener index, Simpson index and Pielou evenness index are used to analyse the diversity of trees and shrubs of the five communities, and the vegetation species diversity of Nanjing Linggu Temple is used for comparison. The indices of the vegetation species diversity of Nanjing Linggu Temple, measured by Professor An Shuqing and Zhao in 1991 (An and Zhao, 1991), are as follows: Shannon–Wiener index 4.562 and Simpson index 0.931, which can represent the vegetation species diversity indices of Purple Mountain. However, the species diversity of each typical section listed in Tables 9.17 and 9.18 is much lower than that of Linggu Temple, especially the Shannon–Wiener index, which is more than twice as low. It is therefore clear that for the plant communities of these sections, there are few species, a lack of species richness and low species diversity.

Table 9.17 Species composition of the five typical sections.

Name	Total number of species	Woody	Broad-leaf	Coni-ferous	Ever-green	Decid-uous	Herba-ceous	Bam-boo
Typical section 1 (K2+600~K3+560)	30	20	18	2	8	12	10	–
Typical section 2 (K40+20~K40+980)	30	19	19	–	8	11	10	1
Typical section 3 (K50+200~K51+175)	32	14	13	1	6	8	16	2
Typical section 4 (K51+175~K52+125)	31	14	13	1	6	8	15	2
Typical section 5 (K60+600~K61+575)	32	22	21	1	9	13	10	–

Table 9.18 Species diversity of the five typical sections.

Name	Shannon–Wiener	Simpson	J_{sw}
Typical section 1 (K2+600~K3+560)	2.114	0.842	0.706
Typical section 2 (K40+20~K40+980)	2.392	0.876	0.798
Typical section 3 (K50+200~K51+175)	1.823	0.777	0.657
Typical section 4 (K51+175~K52+125)	1.118	0.907	0.403
Typical section 5 (K60+600~K61+575)	2.008	0.822	0.649
Linggu Temple (1991) 4.562		0.931	

Although the area of the green belt at both sides of the motorway is small and has a certain impact on the determination of statistics, it can still be seen that the artificial plant community structure is not sufficiently reasonable. For the Pielou evenness index of the typical sections, the species distribution evenness of section 2 is better than that of the rest of the typical sections.

Population life-form: In plant communities, life-forms and structures are closely related. Life-form is the external representation of the adaption of living things to the external environment, and is the result of adaption to climatic conditions, so their composition can reflect the bioclimatic and environmental conditions of a certain region. The life-form system, proposed by the famous Danish ecologist C. Raunkiaer, selects the position of dormant buds during seasons with adverse conditions as the standard for the classification of life-forms, which is widely used due to its simplicity and clarity.

For plant communities growing under the same climatic conditions, the species composition and structure are similar. By comparing the Raunkiaer life-form of artificially planted vegetation communities at typical sections of the Nanjing-Hangzhou Motorway and natural vegetation of adjacent areas (Mount Huangshan), it can be verified whether the construction of the artificiall -planted vegetation communities is scientific and reasonable. The Raunkiaer life-form of Mount Huangshan is: phanerophytes>hemicryptophytes>cryptophytes>therophytes>chamaephytes. From Table 9.19, the life-form of artificially planted vegetation communities along the Nanjing-Hangzhou Motorway is shown as: phanerophytes>chamaephytes> hemicryptophytes>therophytes>cryptophytes for typical sections 1, 2 and 5; and phanerophytes>hemicryptophytes>chamaephytes>therophytes> cryptophytes for typical sections 3 and 4. This is different from the Raunkiaer life-form of the natural vegetation at Mount Huangshan, not reflecting the unique characteristics and features of mixed deciduous and evergreen broadleaf communities of northern subtropical vegetation but rather being similar to the Raunkiaer life-form (phanerophytes> hemicryptophytes>chamaephytes>cryptophytes>therophytes) of temperate evergreen broad-leaf forests on the north slopes of the Qinling Mountains. This indicated species composition and structural arrangement of the artificial vegetation communities at the five typical sections is not reasonable and does not reflect the characteristics of local natural vegetation but rather features more northern warm temperate characteristics. Such kind of plant community is very unstable and prone to frequent change. It is therefore proposed to adopt adjustment and improvement measures to increase the proportion of evergreen tree species and reduce the number of half-shrubs or herbaceous plants.

Table 9.19   Raunkiaer life-form of the five typical sections.

Name	Ph. (%)	Ch. (%)	H. (%)	Cr. (%)	T. (%)	Number of species
Typical section 1 (K2+600~K3+560)	40	26.67	20	3.33	10	30
Typical section 2 (K40+20~K40+980)	36.67	30	20	3.33	10	30
Typical section 3 (K50+200~K51+175)	31.25	18.75	37.5	3.125	9.375	32
Typical section 4 (K51+175~K52+125)	32.26	19.35	35.48	3.23	9.68	31
Typical section 5 (K60+600~K61+575)	43.75	25	18.75	3.125	9.375	32
Mount Huangshan	72.5	1.5	18.1	5.9	2	–
Temperate evergreen broad-leaf forest on the north slopes of the Qinling Mountains	52	5	38	3.7	1.3	–

Ph., phanerophytes; Ch., chamaephytes; H., hemicryptophytes; Cr., cryptophytes; and T., therophytes.

Conclusion

A perfect artificially cultivated plant community should comply with ecological principles, with a reasonable time frame, spatial structure and trophic structure. The trees, shrubs, herbaceous plants and lianas in the community form appropriate seasonal colours. It is important to enrich and compensate for the landscape seasonal defects of the motorway. The selection of tree species within communities should take terrain features, tree species form and ecological characteristics into consideration, so as to establish a well-proportioned, mixed and varied plant structure with rich layers, and present a landscape with frequently changing plant communities. The plant arrangement should fully consider the relationship between species in order to avoid interspecific competition and develop a more stable community system.

After ecological analysis on the vegetation of the five typical sections, it can be concluded that the species selected for the Nanjing-Hangzhou Motorway can generally adapt to site conditions of the area along the motorway and can grow normally. However, there are still some defects, such as low species diversity and lack of a reasonable community structure, with few tree species (especially evergreen broadleaf tree species and coniferous tree species) and many semi-shrubs or herbaceous plants. In addition, the species composition of the five typical sections is extremely similar, resulting in a relatively unvaried appearance of the communities. It is recommended to improve the vegetation structure of these road sections and pay attention to these issues during future landscape design in order to improve the quality of the landscape and design level of the motorway.

10 Vegetation Types and Available Plant Resources along the Nanjing-Hangzhou Motorway

The Nanjing-Hangzhou Motorway passes through Nanjing, Lishui, Liyang and Yixing, with the two main types of natural vegetation being deciduous evergreen broadleaf mixed forest and evergreen broadleaf forest. Plant species adapted to motorways in Jiangsu Province are listed in Appendix 10.A.

10.1 Vegetation in Lishui and Nanjing

Lishui and Nanjing mainly feature deciduous evergreen broadleaf mixed forest. The landscape is mainly hilly from the Ningzhen Mountains on the south bank of the Yangtze River in the north, to the north bank of Lake Taihu via the Maoshan Mountains in the south-east. The south-east of Lake Taihu also features a hilly terrain. The main types of vegetation include:

(1) *Pinus massoniana* – Liquidambar community: for tree layers, the canopy density can reach 0.75, with the constructive species of pine and the average height of *Liquidambar formosana* exceeding 11 m. Associated species include *Ulmus parvifolia*, *Pistacia chinensis*, *Acer buergerianum*, *Celtis sinensis*, *Dalbergia hupeana*, *Sophora xanthantha* and Nanjing *Sabia japonica* Maxim. Shrub layers mainly include *Lindera glauca*, *Symplocos paniculata*, *Rosa multiflora*, *Sageretia thea*, *Serissa foetida* and *Lonicera japonica*. Herbaceous layers mainly include Bryophyta and Pteridophyta, which can grow in sections with relatively dry and barren soils. The vegetation includes the hardy *Pinus massoniana* and colour-changing *Liquidambar formosana*.

(2) *Quercus acutissima* – Liquidambar community: the canopy density of the tree layers is 0.7, with the dominant tree species including *Quercus*

The Environment and Landscape in Motorway Design, First Edition.
Qian Guochao, Tang Shuyu, Zhao Min and Jing Chun.
© 2014 China Communications Press. Published 2014 by John Wiley & Sons, Ltd.

acutissima, Quercus fabri Hance, *Quercus variabilis* and *Liquidambar formosana*, with the height reaching 17 m. The associated species include *Dalbergia hupeana* and *Lindera glauca*. This community type features rich colour changes in autumn, mainly growing in sections with thick, fertile and loose soil.

(3)   *Quercus variabilis – Lindera glauca* community: The canopy density of the tree layers is 0.7, dominated by *Quercus variabilis*, at a height of 15–17 m, with the associated species of *Dalbergia hupeana*. Shrub layers include *Serissa foetida, Symplocos paniculata* and *Lespedeza bicolor* Turcz. Such communities present obvious seasonal changes, being green in summer and golden in late autumn. Mainly grows in sections with thick, fertile and loose soil and can resist infertility to some extent.

(4)   *Liquidambar formosana – Lindera glauca* community: the canopy density of the tree layers is 0.6, dominated by *Liquidambar formosana*, with a height of up to 17 m, with the associated species of *Quercus variabilis, Pistacia chinensis, Quercus fabri* Hance and *Pinus massoniana*. The shrub layers feature rich species, with the main species including *Lindera glauca, Rosa multiflora* and *Symplocos paniculata*, and also including evergreen *Ilex cornuta* Lindl, *Lonicera maackii* and *Serissa foetida*. Such communities present obvious seasonal changes, and grow in sections with favourable soil conditions.

(5)   *Pistacia chinensis – Dalbergia hupeana* community: for tree layers, the canopy density reaches 0.5, dominated by *Pistacia chinensis* and *Dalbergia hupeana*, located at a higher elevation, with a height of up to 10 m. Other associated tree species include *Quercus variabilis, Diospyros lotus* and *Celtis sinensis*. Shrub layers include *Rosa multiflora, Serissa foetida, Lespedeza formosa* and *Lindera angustifolia* Cheng. Such communities feature rich colour changes in autumn, are fond of deep and aerated soil, and are able to endure a certain amount of arid and barren conditions.

(6)   *Quercus variabilis –* Castanopsis sclerophylla community: for tree layers, the canopy density reaches 0.7, dominated by *Pinus massoniana, Quercus variabilis, Castanopsis sclerophylla* and *Liquidambar formosana*, with an average height of up to 10 m. Other associated tree species include *Dalbergia hupeana, Quercus variabilis, Cyclobalanopsis glauca, Ulmus parvifolia* and *Pistacia chinensis*. Shrub layers include *Lindera glauca, Serissa foetida, Rosa multiflora, Ilex cornuta* Lindl and *Acer ginnala* Maxim. Such communities featuring an evergreen landscape in winter are stable local zonal vegetation and are fond of deep and aerated soil.

(7)   *Cyclobalanopsis glauca – Pistacia chinensis* community: for tree layers, the canopy density reaches 0.6, roughly divided into two layers, with the first layer dominated by *Pistacia chinensis* and the second layer dominated by *Cyclobalanopsis glauca*. Other associated species include *Celtis sinensis, Quercus variabilis, Sapindus mukorossi,*

Meliosma cuneifolia Franch, *Liquidambar formosana*, *Acer henryi* Pax, *Sophora xanthantha* and *Ilex chinensis*. Shrub layers include *Diospyros rhombifolia*, *Camellia oleifera*, *Euonymus alatus*, *Ilex cornuta* Lindl and *Acanthopanax gracilistylus*.

(8) *Celtis sinensis – Cyclobalanopsis glauca* community: for tree layers, the canopy density reaches 0.7, roughly divided into two layers according to the height, with the first layer dominated by *Celtis sinensis* and the second layer dominated by *Cyclobalanopsis glauca*. Other associated species include *Phoebe sheareri*, *Pistacia chinensis*, *Liquidambar formosana*, *Juglans cathayensis* Dode, *Hovenia acerba* Lindl, *Acer henryi* Pax and *Sophora xanthantha*. Shrub layers include *Diospyros rhombifolia*, *Camellia oleifera* and *Euonymus alatus*. Such communities featuring an evergreen landscape in winter are fond of deep and aerated soil.

(9) *Phyllostachys edulis* forest: it is unique in terms of community structure, composition, landscape appearance and other aspects, reflecting the subtropical landscape, with no need for deep soil in site conditions. There are also *Phyllostachys glauca* and *Phyllostachys viridis* forest communities, with high ornamental value.

(10) *Lagerstroemia indica* shrub community: *Lagerstroemia indica* grows in relatively poor habitats and is shrubby. Almost pure forests of *Lagerstroemia indica* shrubs exist at the top of Purple Mountain and Maoshan above 300 m, where the soil is arid and barren, and the sunlight is intensive.

In addition, there are also *Spiraea cantoniensis*, *Rhodotypos scandens* and *Exochorda racemosa*, which all have high ornamental value and are adapted to the arid and barren soil of the mountains. These can be used on hills along the route that feature exposed rocks and thin soil layers.

10.2   Vegetation in Liyang and Yixing

Liyang and Yixing are located in the most southern part of Jiangsu Province, featuring the warmest climate and the largest amount of precipitation. In natural plant communities, the quantities and species of evergreen broadleaf tree are significantly higher. The only mid-subtropical broadleaf evergreen forest in Jiangsu Province is mainly distributed across hilly areas of Liyang and Yixing and Dongting Xishan of Lake Taihu. Besides the community types in Lishui, Nanjing, there are also the following unique community types in Liyang and Yixing:

(1) *Cyclobalanopsis glauca – Lithocarpus glabra* evergreen broadleaf plant community: tree layers often include evergreen species such as *Lithocarpus glabra*, *Cyclobalanopsis glauca*, *Castanopsis sclerophylla*, *Machilus*

thunbergii, Myrica rubra, Ilex chinensis, Neolitsea aurata and *Litsea rotundifolia*, as well as deciduous species such as *Quercus glandulifera, Rhus succedanea, Sassafras tsumu,* Tupelo and *Styrax odoratissima*. Shrub layers often include evergreen shrubs such as *Symplocos setchuensis* Brand, *Osmanthus cooperi* and *Ilex fragilis*, as well as deciduous shrubs such as *Rhododendron simsii, Lindera glauca* and *Serissa serissoides*. As the plants of the tree layers and shrub layers grow prolifically, there is a large canopy density and it is therefore dark inside the forest. There are almost no general herbaceous plants, but many sciophilous ferns such as *Rhizoma cibotii. Pteridium aquilinum, Dicranopteris dichotoma, Miscanthus sinensisa* and other light-demanding herbaceous plants grow in gaps and at forest edges. Lianas often include *Smilax china* L. and *Smilax davidiana* A.DC., growing at Yixing Qingshan, Gangxia Zhifang, Tongshan outside Chengze, Dongling Shilongtou and Shengzhuang Xichunling.

(2) *Castanopsis carlesii – Lithocarpus glabra* evergreen broadleaf plant community: this plant community grows at the Longchi Mountain, Yixing. Tree layers often include evergreen species such as *Cyclobalanopsis glauca, Ilex chinensis, Castanopsis sclerophylla, Machilus thunbergii* and *Myrica rubra*, as well as deciduous species such as *Diospyros kaki,* Tupelo, *Sassafras tsumu, Quercus glandulifera, Quercus variabilis, Sophora xanthantha* and *Rhus succedanea*. Shrub layers often include evergreen shrubs such as *Camellia oleifera, Adinandra hirta* Gagnep, *Rhododendron ovatum, Vacciniuim mandarinorum* Diels, *Gardenia jasminoides, Pittosporum sahnianum* Gowda, *Loropetalum chinensis, Eurya muricata, Damnacanthus indicus* and *Elaeagnus pungens,* as well as *Premna microphylla, Ligustrum sinense* Lour, *Rhododendron simsii* and *Viburnum setigerum* Hance. Herbaceous layers often include *Rhizoma cibotii, Parathelypteris glanduligera, Dryopteris fuscipes, Herba lophatheri* and *Polygonum cuspidatum*. Lianas often include *Smilax china* L., *Trachlospermum jasminoides, Caulis piperis* Kadsurae and *Parthenocissus tricuspidata*.

(3) *Phoebe sheareri* evergreen broadleaf plant community: this plant community grows in Langyinjie, Yixing and Jingangjie, Liyang. Tree layers include evergreen species such as *Machilus leptophylla, Litsea rotundifolia* and *Cyclobalanopsis glauca*, as well as deciduous species such as *Semen Hoveniae, Sassafras tsumu, Emmenopterys henryi* Oliv. and *Acer davidii* Franch. Shrub layers are dominated by broadleaf *Indocalamus tessellatus*, and there are also evergreen shrubs such as *Eurya muricata, Camellia sinensis, Rhododendron ovatum, Elaeagnus pungens, Pittosporum sahnianum* Gowda, *Camellia fraternal* and Thymelaeaceae *Daphne,* as well as deciduous shrubs such as *Litsea cubeba, Lindera glauca, Clerodendron cyrtophyllum* and *Broussonetia kazinoki*. As it is damp in the forest, ferns grow prolifically, with common species such as *Arachmiodes simplicior, Dryopteris liyangensis, Rhizoma Polypodiodis Nipponicae* and

Neolepisorus ovatus. Other common herbaceous plants include *Peristrophe japonica, Saxifraga stolonifera, Phryma leptostachya, Pollia japonica* and *Rohdea japonica.*

(4) *Quercus variabilis – Lithocarpus glabra – Rhododendron ovatum* evergreen and deciduous broadleaf mixed communities: tree layers are dominated by *Quercus variabilis* and *Lithocarpus glabra,* including a large proportion of evergreen broadleaf species such as *Castanopsis carlesii, Myrica rubra, Ilex chinensis* and *Cyclobalanopsis glauca,* as well as deciduous broadleaf species such as *Quercus glandulifera, Schoepfia chihehsis* and Tupelo. Shrub layers of this community develop very well, mainly including *Rhododendron ovatum, Rhododendron simsii, Loropetalum chinensis, Vacciniuim mandarinorum, Vaccinium bracteatum, Ardisia japonica, Eurya japonica* Thunb. and *Camellia oleifera.*

(5) *Ilex chinensis – Quercus glandulifera* evergreen and deciduous broadleaf mixed communities: tree layers are dominated by *Ilex chinensis* and *Quercus glandulifera,* and also include Tupelo, *Schoepfia chihehsis, Lithocarpus glabra, Sassafras tsumu, Dalbergia hupeana, Styrax japonicas, Rhus succedanea, Platycarya strobilacea, Albizia julibrissin, Euscaphis japonica* and *Diospyros lotus.* Shrub layers include *Vacciniuim mandarinorum, Lespedeza bicolor, Loropetalum chinensis, Rhododendron simsii, Eurya japonica* Thunb., *Gardenia jasminoides* and *Ardisia japonica.*

(6) *Schima superba – Vaccinium bracteatum* evergreen broadleaf community: tree layers are dominated by *Schima superba,* and also include *Quercus glandulifera, Myrica rubra, Castanopsis sclerophylla, Quercus fabri* Hance and *Symplocos setchuensis* Brand. Shrub layers are dominated by *Vaccinium bracteatum,* and also include *Rhododendron, Phyllostachys glauca, Eurya japonica* Thunb. and *Rhamnus crenata.*

(7) *Castanopsis sclerophylla – Phoebe sheareri* community: the species composition of the tree layers is abundant, including *Castanopsis sclerophylla, Phoebe sheareri, Ilex chinensis, Quercus variabilis, Liquidambar formosana, Zelkova schneideriana, Meliosma cuneifolia, Hovenia acerba, Acer buergerianum* and *Cyclobalanopsis glauca.* Shrub species mainly include *Euonymus alatus, Phyllostachys glauca, Ardisia japonica* and *Rosa multiflora.*

Coniferous forest is dominated by *Pinus massoniana* forest and *Cunninghamia lanceolata* forest, with the former generally identical with that at Dongting Mountain in Suzhou. However, mixed broadleaf species, especially evergreen species, are slightly more in terms of types and number. For example, in the *Cunninghamia lanceolata* forest at Mingling Mountain in Yixing there are many mixed broadleaf species due to seldom conservation and management. The tree layer includes species such as *Ilex chinensis, Rhus chinensis, Fraxinus chinensis, Sassafras tsumu, Myrica rubra, Litsea rotundifolia* and *Styrax odoratissima,* and the shrub layer includes species such as *Eurya muricata, Azolla imbricate, Rhododendron ovatum, Lindera glauca, Vacciniuim mandarinorum, Vaccinium bracteatum, Indigofera fortune, Viburnum setigerum* Hance, *Viburnum erosum* Thunb., *Rhododendron molle, Glochidion puberum, Kerria japonica, Eurya hebeclados* and Thymelaeaceae *Daphne*. Herbaceous plants often include *Deyeuxia arundinacea, Miscanthus sinensis* and *Imperata cylindrical*, followed by *Solidago virgaurea, Eupatorium japonicum* and *Kalimeris indica*.

10.3   Exploitable plant resources

Vegetation types and structure

Vegetation type varies with the natural environmental conditions and every vegetation type has its own morphology and structural features. There is deciduous and evergreen broad-leaved mixed forest and evergreen broad-leaved forest along the Nanjing-Hangzhou Motorway, which are composed of tree layers, shrub layers and herbaceous layers. However, different coenotypes have different colonizing plants and dominant plants. In the community, different layers of plants live together due to their complementarity in ecological character, and they build a stable colony system and display their own macroscopic features, such as morphology, height and colour. When the natural vegetation is subject to natural or man-made damage such as cutting a mountain or slope, people have to imitate, reconstruct or reproduce the damaged vegetation by referring to the structural features and morphology of the native vegetation. The vegetation type is therefore an indispensable landscape element needed to recover and create the regional environmental landscape, and also a 'blueprint' to restore the regional vegetation. For this reason, they should be studied and utilized carefully.

Species of wild plants

A region's wild plants and indigenous plants are good materials for local environmental greening because of their strong adaptability, rapid growth, strong stress resistance and distinct regional characteristics. They also do not have the ecological safety problems of foreign species. They are especially useful in motorway construction, where indigenous plants can increase the diversity of greening plants and enrich the landscape of greening plants along the route. With more and more attention being paid to ecological landscape construction, constantly exploring local wild plant resources and gradually exploiting them is also an important task to integrate the whole environment of the motorway with nature and ensure coordination with the environment. The prospects for this are bright. According to investigation, the wild flowering shrubs and tree species available for the Nanjing-Hangzhou Motorway mainly include those given in the following.

Evergreen broad-leaved trees

Castanopsis sclerophylla: a Fagaceae evergreen tree with a spherical crown. Its natural distribution stretches as far north as the Purple Mountain in Nanjing. It has the strongest cold endurance capacity among all evergreen broad-leaved trees and is also a dominant species in the zonal climax of this area. It is suited to growing in moist aerated soil. Its branches and leaves are exuberant and prune-enduring, and it is appropriate to plant in service areas or building areas to provide shade. It can also be used as a background tree. It may be independently planted on grass next to interchanges or on open ground, or may be planted in blocks and groups as a colonizing species of evergreen broad-leaved forest. It is also suitable to plant as landscape forest on slopes or hilly land beside the motorway (Figure 10.1).

Lithocarpus glabra: a Fagaceae *Lithocarpus* evergreen tree with a hemispherical crown. It is distributed in all provinces and regions south of the Yangtze River, often in mountainous and hilly regions below 500 m above sea level. It is inherently a frigostable, sciophilous species. It is also thermophilous and fond of damp and thick soil, as well as drought-enduring and barren-enduring.

Figure 10.1 Castanopsis sclerophylla.

Its branches and leaves are exuberant and dark green. It may be planted alone in service areas and building areas to provide shade, or can planted in blocks and groups as an ornamental forest in the interchange area together with *Castanopsis sclerophylla*, *Quercus serrate*, *Quercus fabri*, *Liquidambar formosana*, *Pinus massoniana* Lamb and *Quercus variabilis*, in order to form a zonal plant community in Jiangsu Province.

Phoebe sheareri: a Lauraceae *Machilus* evergreen tree. It can be found in areas below approximately 300 m above sea level, in the south of Jiangsu Province and also grows in Nanjing. It is thermophilous and fond of a moist climate, as well as thick, fertile and damp slightly acidic and neutral soil with good dewatering. It is frigostable, deep-rooted and features strong germination but it grows slowly. Its form is regular and attractive and its leaves are big, providing a large area of shade. If it is planted with other evergreen trees as a background forest, the results are even better.

Cyclobalanopsis glauca: a Fagaceae evergreen tree. It grows in areas below 1000–1600 m above sea level in vertical distribution. It is thermophilous and fond of a rainy climate as well as sciophilous. It is also fond of calcium-rich soil, so it is often seen in limestone mountainous areas. It also grows well in acidic soil with good dewatering and thick humus. This tree has dense branches and leaves and an attractive form. It retains its leaves all year round. Because it is sciophilous, it is an important tree species forming the evergreen broad-leaved forest in the south of Jiangsu Province. It is suitable to plant in interchange areas, at the edge of grass land and in service areas and building areas. It is can also be planted with *Schima superba*, *Elaeocarpus sylvestris* and *Liquidambar formosana* to form an evergreen broad-leaved deciduous forest community, or may be planted as a hedgerow to resist wind, fire and noise (Figure 10.2).

Figure 10.2   Cyclobalanopsis glauca.

Cyclobalanopsis myrsinaefolia: it is distinguished from *Cyclobalanopsis glauca* mainly by its narrow leaves (1.5–2.5 cm wide) with tiny sawtoothed edge. It is distributed in provinces and regions south of the Yangtze River and naturally grows in mountainous areas in Yixing and Liyang. It grows well in the open in Nanjing. It is suitable for fertile and loose soil, and can also grow in the cracks of rocks. Its landscaping effect and use are the same as for *Cyclobalanopsis glauca*.

Schima superba: a Theaceae evergreen tree with an oval crown. It is fond of a warm, hot and moist climate, and well-adapted to soil and deep-rooted. It grows at a moderate speed and is often seen in valleys and forest land 150–1500 m above sea level. Because the fully grown tree is tall, big and photophilous, it is often the upper canopy species in the community. There is a wild population distributed in Guangfu on the Taihu Lake section located in Suzhou. It may be planted at sections in Yixing and Liyang. It not only has an attractive form, but it is also not susceptible to burning, so it may be often used as a species in the fire-barrier belt.

Photinia davidsoniae: a Rosaceae evergreen tree. Its trunk has multi-branches in clusters from its stem base, with luxuriant foliage and shiny leathery leaves. It has strong resilience to cold weather and good adaptivity, so it may be planted at both sides of the motorway. With dense branches and leaves, a shiny green colour and elegant form, it is suitable to plant in interchange areas, service areas, building areas, and outside the slope and ditch. It may be planted alone, in rows or together with other plants beside rocks, roads and grass. It is often used as a background tree for plants with bright-coloured flowers or light-coloured leaves; it may also be densely planted as high hedgerow (Figure 10.3).

Xylosma racemosum: a kind of Flacourtiaceae evergreen dunga-runga. Its trunk has multi-branches in clusters from its stem base, with luxuriant foliage, thorns on its branches, and shiny leathery leaves. There is wild distribution in Nanjing. It has strong resistance to cold weather, is capable of growing well in deep and fertile soil, and is also resilient to a certain degree of barren and thin soil. With strong adaptability and prune-enduring properties, it is suited to make hedgerows and barrier belts. It is therefore often arranged on slopes and outside ditches on both sides of the motorway (Figure 10.4).

Ilex chinensis: an Aquifoliaceae evergreen dunga-runga with dense branches and leaves and regular tree form. Its bark is dark bluish green and smooth and it has thin leathery leaves. This species is photophilous and sciophilous. It is also fond of a warm and humid climate and fertile and acidic soil. It is humidity-enduring, but not cold-resistant. It has a strong sprouting ability, is prune-enduring, deep-rooted and resistant to wind, and grows slowly. It has exuberant and evergreen branches and has many attractive red fruits which it retains through the winter, so it is therefore suitable for planting in interchange areas, service areas and building areas as a roadside tree and landscape tree. It may be planted alone or in groups on grass, in front of buildings, next to lakes and at the edge of forests. It may be used as a hedgerow or arranged with

Figure 10.3   **Photinia davidsoniae.**

Figure 10.4   **Xylosma racemosum.**

other ornamental plants. It can also be planted alone or in groups within the interchange (Figure 10.5).

Myrica rubra: a Myricaceae evergreen dunga-runga with a neat and approximately spherical crown. It is mainly distributed in provinces and regions south of the Yangtze River. It is natural and slightly sciophilous, and cannot endure direct sunlight. It is fond of a warm and humid climate and acidic soil with good dewatering; it may also grow in neutral and subalkaline soil. With its exuberant branches and leaves and round crown, it may be planted alone in service areas and building areas or in groups at the edge of grass, around buildings or in a row at the edge of a road. It may also be densely planted within the interchange to isolate spaces or shade the pavement (Figure 10.6).

Michelia chapensis: a Magnoliaceae *Michelia* evergreen tree with long oval crown and thin leathery leaves. It is distributed in red soil or yellow soil in mountainous land, and often grows in the thick-soil sections of medium and lower slopes. It is fond of moisture and not resilient to waterlogged environments. It is a mesophilous tree species. With a straight and tall trunk, dense branches and leaves and an attractive tree form, it may be planted alone or in groups at both sides of the motorway as roadside trees or greening trees in gardens.

Figure 10.5   Ilex chinensis.

Figure 10.6 Myrica rubra.

Parakmeria lotungensis: a Magnoliaceae evergreen tall tree with narrow obovate leathery leaves and flowering phase in April and May. It has been introduced to the Nanjing region and grows well without freeze injury. It is fond of a humid and warm climate. As a mesophilous tree species, it often grows in slightly acidic or neutral fertile loose soil with good air permeability. This species is tall and straight from top to bottom with sparse branches and shiny leaves. With aubergine and tawny tender leaves, it is a valuable greening tree species. It may be used as greenery in service areas and sections with good water and soil conditions.

Elaeocarpus sylvestris: an Elaeocarpaceae evergreen tree with an ovoid crown, puce smooth indehiscent bark and thin leathery leaves. Its leaves are shaped like a long obovate oval, and it features axillary-racemes and white petals. It has been introduced to Nanjing and Wuxi and grows well. It is fond of a moist and warm climate with some resilience to cold conditions. It is suited to growing in yellow soil and yellowish red soil with good dewatering. This species has exuberant branches and leaves with a round and neat crown. After the frost season, some leaves become red and green, making for a very beautiful scene. It is suitable for planting in interchange areas, service areas and building areas; it may also be used as a roadside tree and garden tree. It is often planted in groups on grass, at intersections and at the edge of the forest; it may also be planted in rows on slopes or outside ditches to protect the slope, shade the road or act as a noise barrier (Figure 10.7).

Deciduous broad-leaved trees

Quercus acutissima: a Fagaceae deciduous megaphanerophyte with crisscrossing bark featuring deep cracks and small tawny branches. It is photophilous, cold-resistant, drought-enduring and fond of a humid climate. It does not have high requirements for soil quality but cannot endure saline-alkaline soil. It grows well in mountainous and hilly land. It is deep-rooted, has a strong sprouting ability and medium growth speed. Its trunk is straight from top to

Figure 10.7   **Elaeocarpus sylvestris.**

bottom with extensile branches and an attractive form. In summer, its leaf colour is dark green and in autumn it becomes orange. It is suitable for planting with other tree species to form a mixed forest. It has strong resilience to wind, fire and smoke, meaning that it is suitable for ornamental forests and protection forests.

Quercus variabilis: a Fagaceae deciduous megaphanerophyte with a widely ovate crown, taupe bark and deep split cracks on the bark. It is photophilous and often grows on sunny slopes in mountainous land. However, its sapling prefers to grow in shade. It has strong adaptability to climate and soil, and is suited to growing in humid and sandy soil with good dewatering; it cannot endure waterlogged environments. It has a towering trunk with a wide and extensile crown, high branches, rough bark and a changing leaf colour, resembling *Quercus acutissima*. It is suitable for planting in groups in mountainous land, sloping valleys or together with other tree species. Because its root system is very developed with strong adaptability and its thick leaves are not prone to burning, thus it serves as an excellent species for windbreak forests, water conservation forests and fire prevention forests.

Sophora xanthantha: a Leguminosae deciduous dunga-runga. It grows at the side or edge of a road or in sandy soil on the hillside. It is photophilous and grows well in moist site conditions. It can endure drought, barren and thin soil and dry air and grow along the road. Its crown spreads out like an umbrella with fresh and unique leaves, so it can be used as a shade tree and roadside tree.

Pistacia chinensis: an Anacardiaceae deciduous tree with a nearly spherical crown (Figure 10.8). As a pioneer naturally growing species in barren hills, it is often scattered amongst low mountains, hills and flat areas. It is photophilous and its sapling is slightly sciophilous. It is fond of a warm climate and not resilient to severe cold. It is also drought-enduring and barren-enduring. It docs not have high requirements for soil, so it grows well in slightly acidic, neutral and subalkaline sandy and clay soil. Its crown is perfectly round with exuberant beautiful branches and leaves. In early spring, its tender leaves are red; in autumn, it becomes dark red or orange-yellow. Its red pistillate inflorescences are very attractive, suitable for planting in interchange areas, service areas and building areas as a shade tree, roadside tree and coloured-leaf species in autumn. It may be planted in groups or with *Liquidambar formosana*, *Ginkgo biloba* L. and *Sapium sebiferum*, so as to form a beautiful autumnal scene. This tree is deep-rooted and has well-developed main roots with a strong wind-proof capacity, so it may be planted on slope and outside ditches as a species for consolidating the soil and protecting forests.

Celtis sinensis: an Ulmaceae deciduous tree with an oblate crown. It is photophilous and slightly sciophilous. It is fond of a warm climate and fertile, moist, thick neutral clay soil, and can also endure slightly saline-alkaline soil. It is deep-rooted with a strong wind-proof capacity. Its form is beautiful with

Figure 10.8   Pistacia chinensis.

a broad crown and dark green shade; it can be arranged in sections not prone to waterlogging with thick soil (Figure 10.9).

Aphananthe aspera: an Ulmaceae deciduous tree with a spherical crown. There is wild distribution in Jiangsu. It is photophilous and fond of a warm and moist climate. It is suited to growing in humid, fertile and thick acidic soil. Its trunk is tall and straight with a wide crown and exuberant branches and leaves. It can endure some waterlogging, so it may be planted beside roads and streams.

Pseudolarix amabilis: a Pinaceae deciduous tree endemic to China with a widely conical crown. It is one of the most famous ornamental trees in the world. Because its leaves on the brachyplast are fascicles like copper coins and become golden in late autumn, it is called the 'golden pine' (Figure 10.10). Its form is tall and magnificent, and its leaves and crown are colourful, meaning that it is elegant and pleasing to the eye. When they are arranged symmetrically, if broad-leaved trees and evergreen shrubs are planted with them, it

Figure 10.9   Celtis sinensis.

results in a dazzlingly beautiful and varied landscape with distinct layers. *Pseudolarix amabilis* often grows with *Ginkgo biloba* L., *Cryptomeria japonica* var. *sinensis* and *Liquidambar formosana* to form a beautiful natural scene at the Tianmu Mountain west of Zhejiang Province. It is suitable for planting in rows or in blocks in interchange areas, service areas, near buildings, on slopes and outside ditches and at the edges of open grass; alternatively, it can be planted with coloured-leafed trees (such as *Liquidambar formosana* and *Acer palmatum*) and evergreen species to create colourful autumn scenery.

Diospyros lotus: an Ebenaceae deciduous tree. It is photophilous, semi-sciophilous and moisture-proof. It is fond of fertile and thick soil but it can also endure barren, medium alkaline soil and calcareous soil to a certain degree. Its trunk is straight and upright with a round and neat crown and strong adaptability. It may be planted in all kinds of site conditions at both sides of the motorway.

Sassafras tsumu: a Lauraceae deciduous tree with a widely oval or spherical crown. It is photophilous and not shade-enduring and fond of a warm and humid climate and thick acidic soil with good dewatering. Its trunk is

Figure 10.10   Pseudolarix amabilis (in autumn).

straight from top to bottom with wide and unique leaves. In late autumn, its leaves become reddish yellow; in spring, small yellow flowers blossom over the leaves, creating a very beautiful scene.

Cornus controversa: a Cornaceae deciduous tree. It is photophilous and slightly shade-enduring. It is fond of a warm and humid climate and resilient to a certain degree of cold. It prefers fertile and moist soil with good dewatering. Its form is neat and regular with its large lateral branches resembling a lampstand, forming a beautiful conical crown.

Alnus trabeculosa: a Betulaceae Alder deciduous tree or shrub. It grows in places such as Yixing, Jurong and Suzhou. It is often found beside brooks.

Phellodendron amurense: a Rutaceae deciduous tree with a broad crown. It is photophilous and cold-resistant, but not shade-enduring. It is fond of suitably moist neutral or slightly acidic soil with good dewatering. It does not grow well in clay soil or in barren soil. It is deep-rooted with a strong wind-proof capacity. Its crown is wide and sparse. In autumn, its leaves become a very beautiful yellow. It may therefore be planted at both sides of the motorway.

Evergreen shrubs

Camellia oleifera: a Theaceae evergreen tree or shrub. It is mainly distributed in the Yangtze River basin and provinces and regions south of the River. It is fond of a warm and humid climate and is photophilous. With its sapling being shade-enduring, it is a shade-intolerant subtropical species. It is suitable for planting in groves, at the edge of forests and at the corners of roadsides. It is often planted in clusters or in groups, and sometimes alone (Figure 10.11).

Vaccinium bracteatum: an Ericaceae evergreen shrub growing in the bush fallow shrubland on the hillsides or in undergrowth. It is suited to growing in the acidic soil sections in Yixing and Liyang.

Rhododendron ovatum: an Ericaceae evergreen shrub with smooth glabrous branches and leaves. Its leaves are leathery and oval with a sharp or blunt apex and round base; its flowers are monotonous and grow between the axils on top of its branches. It is fond of acidic soil and intolerant of alkaline and clay soil. Due to its fondness of sunlight and a cool and moist climate, it is most suitable for planting in groups in the moist and shady undergrowth of forests and between rocks. It is widely distributed in mountains and plains. When in bloom, its flowers bundle together like brocades, turning countless mountain

Figure 10.11 Camellia oleifera.

slopes into a ubiquitous red. It is suited to growing in acidic soil sections beside roads and under the deciduous forest.

Adinandra hirta Gagnep: a Theaceae or shrub or small tree. It grows on hill-sides or in forests in valleys. It is distributed in provinces south of the Yangtze River.

Elaeagnus pungens: an Elaeagnaceae *Elaeagnus* evergreen shrub. It is distributed in provinces south of the Yangtze River. It is photophilous and semi-sciophilous, and fond of a warm climate and not cold-resistant. It has a strong adaptability to soil, enduring both drought and water-dampness. It features crisscrossing branches, silvery grey backs to its leaf blades, fragrant flowers and brightly coloured fruits, giving it a high ornamental value (Figure 10.12).

Deciduous shrubs

Spiraea cantoniensis: a Rosaceae deciduous shrub with tall and slender, arched and glabrous branches. Its leaves are a diamond-oblong to diamond-lanceolate shape with a deep-cut sawtoothed edge; both sides are smooth with a dark green surface, turquoise back and wedge-shaped base. Its inflorescence is umbrella-type and smooth. This species is photophilous and fond of a warm and humid environment, with its cold-resistance inferior to other species of

Figure 10.12   Elaeagnus pungens.

Figure 10.13 Spiraea cantoniensis.

the same genus. Its branches and leaves are dense and exuberant. It blossoms with dense inflorescence in late spring. It has dense pure white flowers, making it an excellent ornamental shrub (Figure 10.13).

Crataegi cuneatae: a Rosaceae deciduous shrub that sometimes looks like a tree. Its branches are dense and thorny. Its tender branches have pubescence, but its grown branches are glabrous. Its blades are widely obovate or obovate and oval with its base being wedge-shaped; its decurrence connects with the petiole and irregular double saw tooth at the edge. Its apex often has 3 or sometimes 5–7 shallow lobes with the vein on the surface having pubescence which then sheds later. The back of the blade and petiole feature sparse pubescence; its stipule is sickle-shaped with a big saw tooth. There are 3–7 flowers on its corymb; there is pubescence on its peduncle and anthocaulus, and its flowers are white. It grows throughout Jiangsu Province; it grows in the wild in mountainous shrubland.

Lespedeza formosa: a Leguminosae deciduous shrub. It is common in the wild in many regions and grows in the undergrowth of hillside forests or amongst weeds. It may be used as a slope protection plant (Figure 10.14).

Styrax faberi: a Styracaceae deciduous shrub or small tree. With beautiful flowers and fruits, it can be used as an ornamental species.

Euonymus alatus: a Celastraceae deciduous shrub. It is photophilous and slightly shade-enduring. It has strong adaptability to climate and soil. Able to endure drought, barren and cold conditions, it can be grown in neutral, acidic and calcareous soil. Its branches are unique. In early spring, it sprouts tender leaves and in autumn, its leaves are a very attractive purplish-red. After defoliation, there are purple small fruits on the branches, which are very attractive, so it is an excellent ornamental fruit tree, and especially unique as a hedgerow plant.

Figure 10.14 Lespedeza Formosa.

Lindera angustifolia Cheng: a Lauraceae deciduous shrub or small tree. It is distributed in mountainous and hilly regions in the south of Jiangsu Province; and grows in shrubland on the hillside. It may be used as a slope protection plant.

Viburnum macrocephalum: a Caprifoliaceae deciduous shrub. It is fond of thick and fertile sandy soil with good dewatering but does not enduring waterlogging. It is short with a delicate and pretty appearance. It blossoms in spring with beautiful and unique flowers; in late autumn, it produces attractive vivid red fruits.

Diospyros rhombifolia: an Ebenaceae deciduous shrub. It is a subtropical species. It is cold-resistant and photophilous and often grows in shrubland in valleys exposed to the sun. It does not place high demands on the soil, so it grows well in acidic and neutral soil, as well as in limestone mountainous land. It can endure dry and barren conditions. Its root system is very developed. In autumn, the shrub gets red leaves.

Exochorda racemosa: a Rosaceae deciduous shrub. It is often found in shrubland in the gravel soil on low hilly land. It can grow in acidic and neutral soil; and can endure dry and barren conditions. After it is introduced to plain areas, it grows exuberantly in soil with good dewatering, and possesses a strong cold-resistant capacity. It has beautiful white flowers, so it is often used as an ornamental species. It is suitable for planting on grass, at the edge of forests, at roadsides and in a rockery. If it is planted in groups at the edge of evergreen groves, it creates a very elegant scene, looking like flakes of snow amongst the woods.

Caragana sinica: a Leguminosae deciduous shrub. This species is photophilous, cold-resistant and has strong adaptability. It does not place high demands on the soil and can endure drought and barren conditions, as well as being able to grow in rocky cracks. It has beautiful bright-green leaves. It may be planted next to rocks and at roadsides or used as a hedgerow, as well as a slope protection plant.

Weigela coraeensis: a Caprifoliaceae *Weigela* deciduous shrub as tall as 5 m when fully grown. It is photophilous and slightly shade-enduring, and fond of moist and fertile soil.

Woody climbing plants

Rosa multiflora: a Rosaceae deciduous shrub with long stems. It is decumbent or creeping with thorny stipules. Its flowers are white or baby pink and fragrant. It has good adaptability and is photophilous, cold-resistant and does not place high demands on the soil. It can grow normally in heavy soil, most suitable for planting as a flower hedge. It is also good to plant in clusters on hillsides to conserve water and soil.

Rosa cymosa: a Rosaceae deciduous trailing shrub with hook-like prickles. It is a floribunda with its flowering phase in April and May. Its fruiting time is from August to October. It is distributed throughout Jiangsu Province, and grows in the wild on mountain slopes or on hills.

Lonicera japonica: a Caprifoliaceae semi-evergreen twining shrub. It is photophilous and shade-enduring, cold-resistant, drought-enduring and dampness-enduring. It does not place high demands on the soil and can grow in acidic and alkaline soil. It has high robustness and adaptability and a developed root system. Its plant has a tender appearance with intertwining cirri. In winter, its leaves are slightly red. It can be planted on the rocky hills and beside ditches near streams. It is also suitable for planting on hillsides as a ground cover plant.

Lardiza balaceae: a Lardizabalaceae deciduous vine without any pubescence. It has palmately compound leaves with 5 leaflets in obovate or oval shape. Its leave apex is blunt or slightly sunken with an entire edge, and grows light-purple flowers. It is slightly shade-enduring and fond of a warm and moist climate as well as soil with good dewatering. It is often found in open forest, on hillsides or beside paddy fields. Its viticulas intertwine with a unique leaf shape. Its flowers are pulpy and purple with 3–5 forming one cluster. It is a vine with beautiful flowers and leaves (Figure 10.15).

Figure 10.15   Lardiza balaceae.

Perennial herbs

Reineckia carnea: a Liliaceae perennial evergreen herb with spherical red berries. Its flowering phase is from October to November. It may be used as a ground cover plant (Figure 10.16).

Asparagus umbellatus: a Liliaceae perennial evergreen herb with its flowering phase in May and fruiting time in August. It is distributed in Jiangsu Province and grows on hillsides or at riversides. It is also distributed in Hebei, Shanxi, Shaanxi and Gansu and provinces and regions in East China, Central Southern China and Southwest China.

Figure 10.16   Reineckia carnea.

Hosta plantaginea Aschers: a Liliaceae perennial fascicular herb with white, horny fragrant flowers. Its filament base is concrescent with its chlamydeous tube and its capsules are cylindrical or prismoid, 4.5 – 7 cm in length. Its flowering phase is in August and September and fruiting time is in September and October. It originates from Jiangsu Province. It is now cultivated all over the country for viewing. It is suited to growing in dark and moist areas.

Slope soil consolidating plants

Zoysia sinica Hance: a Gramineae perennial low herb 10 – 30 cm in stem height without pubescence on its leaf sheath but with pubescence on the sheath opening. It grows throughout Jiangsu Province, usually on river banks and the edge of roads. It is distributed in areas such as North China, East China and Hainan Island. It is a favourable variety for turfing.

Hemerocallis fulva: a Liliaceae perennial fasciculate herb with fleshy cambiform inflated tuberous root underground. Its basal leaves are pale green with a wide linear margin and high stems. There are 6 – 10 jacinth-coloured flowers on its inflorescence with a flowering phase from June to August. It grows all over Jiangsu Province, and can be found in the wild next to valleys or dark and humid areas under the forest. It is now cultivated throughout China.

Viola tricolor var. *hortensis*: an Iridaceae perennial herb with thick vertical rhizomes and slender horizontal rhizomes. Its blades are thick, shiny and ensiform. Its flowers are lavender or blue with a light yellow base. Its flowering phase is in June and July. It grows in the south of Jiangsu Province.

Forsythia suspense: an Oleaceae deciduous shrub with a fasciculate upright stem and extensile arched down-growing branches. Its branchlets are tawny and nearly quadrangular with obvious lenticels and hollow pith. Its leaves are opposite and oval with a sharp apex. It is photophilous, shade-enduring to a certain degree, cold-resistant, drought-enduring and barren-enduring but not resilient to waterlogging. It does not place high demands on the soil and is strongly resilient to pests. Its branches are arched and extend outwards. In early spring, its flowers bloom earlier than its leaves, with its flowering phase in April and May, dressing the branches all over with a beautiful golden colour. Its root system is well developed, which protects the embankment (Figure 10.17).

Bromus inermis: a Gramineae perennial herb with horizontal rhizomes and upright stem, which usually has no pubescence. Its flowering phase and fruiting time are from April to August. It often grows on hillsides, the edge of roads and river banks. For its horizontal rhizomes, it may be a pioneer plant for consolidating sand and soil. It is therefore suitable to use as a biological protection herb on motorway slopes.

Agrostis stolonifera: a Gramineae perennial thin and delicate herb with thin and delicate rhizomes and delicate stem 20 – 50 cm in height. Its flowering phase and fruiting time are from April to July. It grows throughout Jiangsu Province, often found on hillsides, fields and humid areas. It is also found in provinces ranging from East China via Central China to South China and Southwest China.

Figure 10.17   Forsythia suspense.

Appendix 10.A   Plant Species Adapted to Motorways in Jiangsu Province

Evergreen trees

Cedrus deodara
(2)Pinus bungeana
(3)Pinus elliottii
(4)Pinus parviflora
(5)Pinus taeda
(6)Pinus thunbergii
(7)Cryptomeria japonica var. sinensis
(8)Juniperus formosana
(9)Platycladus orientalis
(10)Sabina chinensis
(11)Sabina chinensis var. kaizuca
(12)Sabina virginiana
(13)Podocarpus macrophyllus
(14)Myrica rubra
(15)Castanopsis sclerophylla
(16)Cyclobalanopsis glauca
(17)Lithocarpus glabra

(18)*Nandina domestica*
(19)*Magnolia grandiflora*
(20)*Manglietia fordiana*
(21)*Michelia figo*
(22)*Michelia maudiae*
(23)*Ternstroemia gymnanthera*
(24)*Cinnamomum camphora*
(25)*Pittosporum tobira*
(26)*Distylium racemosum*
(27)*Loropetalum chinense var. rubrum*
(28)*Eriobotrya japonica*
(29)*Photinia davidsoniae*
(30)*Photinia serrulata*
(31)*Pyracantha fortuneana*
(32)*Ilex chinensis*
(33)*Ilex cornuta* Lindl
(34)*Ilex crenata*
(35)*Ilex latifolia*
(36)*Euonymus japonicus*
(37)*Camellia japonica*
(38)*Camellia oleifera*
(39)*Camellia sasanqua*
(40)*Elaeagnus pungens*
(41)*Ligustrum lucidum*
(42)*Ligustrum quihoui*
(43)*Osmanthus fragrans*
(44)*Nerium oleander*
(45)*Gardenia jasminoides*
(46)*Viburnum odoratissimum*
(47)*Citrus reticulata*
(48)*Buxus harlandii*
(49)*Buxus sinica*
(50)*Elaeocarpus decipiens*
(51)*Phyllostachys propingue*
(52)*Mahonia fortunei*
(53)*Xylosma racemosum*

Deciduous trees

(1)*Pseudolarix amabilis*
(2)*Metasequoia glyptostroboides*
(3)*Taxodium ascendens*
(4)*Taxodium distichum*
(5)*Populus tomentosa*
(6)*Salix babylonica*

(7)Salix chaenomeloides
(8)Salix matsudana
(9)Pterocarya stenoptera
(10)Celtis sinensis
(11)Ulmus parvifolia
(12)Zelkova schneideriana
(13)Broussonetia papyrifera
(14)Ficus carica
(15)Morus alba
(16)Liriodendron chinense
(17)Magnolia denudata
(18)Magnolia liliflora
(19)Magnolia zenii
(20)Chimonanthus praecox
(21)Liquidambar formosana
(22)Amygdalus persica
(23)Armeniaca mume
(24)Cerasus serrulata
(25)Chaenomeles sinensis
(26)Chaenomeles speciosa
(27)Malus halliana
(28)Spiraea cantoniensis
(29)Spiraea prunifolia
(30)Albizia julibrissin
(31)Cercis chinensis
(32)Amorpha fruticosa
(33)Pistacia chinensis
(34)Rhus typhina
(35)Rosa roxburghii
(36)Euonymus maackii
(37)Acer buergerianum
(38)Acer palmatum
(39)Aesculus chinensis
(40)Koelreuteria integrifoliola
(41)Sapindus mukorossi
(42)Hibiscus mutabilis
(43)Hibiscus syriacus
(44)Firmiana simplex
(45)Edgeworthia chrysantha
(46)Lagerstroemia indica
(47)Punica granatum
(48)Camptotheca acuminata
(49)Cornus alba
(50)Rhododendron simsii
(51)Rhododendron mucronatum

(52)*Diospyros kaki*
(53)*Sinojackia xylocarpa*
(54)*Forsythiaviridissima*
(55)*Paulownia tomentosa*
(56)*Lonicera maackii*
(57)*Viburnum macrocephalum*
(58)*Catalpa bungei*
(59)*Robinia pseudoacacia*
(60)*Sophora japonica*
(61)*Wisteria sinensis*
(62)*Zanthoxylum bungeanum*
(63)*Ailanthus altissima*
(64)*Melia azedarach.*
(65)*Alchornea davidii*
(66)*Bischofia polycarpa*
(67)*Sapium sebiferum*
(68)*Platanus acerifolia*
(69)*Vitis vinifera*
(70)*Actinidia chinensis*
(71)*Syringa vulgaris*
(72)*Salix × aureo-pendula var. vitellina*
(73)*Carya illinoensis*
(74)*Rhus chinensis*

Grass and ground cover

(1)*Cynodorz ctylorz*
(2)*Cynodon dactylon × C. transvaalensis*
(3)*Zoysia matrella*
(4)*Zoysia tenuifolia*
(5)*Festuca arundinacean*
(6)*Poa praterzsis*
(7)*Lolium perenne*
(8)*Trifolium repens*
(9)*Dichondm repens*
(10)*Ophiopogon japonicus*
(11)*Zephyranthes canndida*
(12)*Coreopsis lanceolata*
(13)*Coronilla varia*
(14)*Dendramthema nankingense*
(15)*Dianthus plumarius*
(16)*Hemerocallis fulva*
(17)*Iris tectorum*
(18)*Liriope platyhylla*
(19)*Lycoris radiate*

(20)*Oxalis rubra*

(21)*Orychophraqmus violaceus*

(22)*Reineckia carnea*

(23)*Acorus calamus*

(24)*Alisma orientale*

(25)*Sagittaria sagittifolia*

(26)*Quamoclit pennata*

(27)*Trachlospermum jasminoides*

(28)*Vinca major*

(29)*Campsis grandiflora*

(30)*Rosa chinensis*

(31)*Kerria japonica*

(32)*Lespedeza formosa*

(33)*Millettia reticulata*

(34)*Parthenosissus tricuspidata*

(35)*Hypericum monogynum*

(36)*Fatsia japonica*

(37)*Aucuba japonica* 'Variegata'

(38)*Jasminum mesnyi*

(39)*Ligustrum ovalifolium Vicaryi*

(40)*Euonymus fortunei*

(41)*Berberis thunbergii* 'atropurpurea'

(42)*Shibataea chinensis*

(43)*Hedera nepalensis*

(44)*Sasa fortunei*

(45)*Medicago sativa*

11 Green Landscape Design of the Nanjing-Hangzhou Motorway (Phase II)

11.1 Project overview

The completion and opening to traffic of the Nanjing-Hangzhou Motorway (Phase I, from Lishui to Yixing in Jiangsu Province) is a pioneering event in terms of 'nature, environmental protection, tourism and scenery' motorways in Jiangsu Province, serving as a model for motorway construction in China. For construction of the Nanjing-Hangzhou Motorway (Phase II), the construction concept of 'nature, environmental protection, tourism and scenery' was strengthened and greater efforts were made to ensure environmental protection, ecological maintenance, landscape construction and people-oriented facilities, maximize the overall effectiveness of the green landscape, and speed up the connection between Nanjing and Hangzhou.

Phase II of the Nanjing-Hangzhou Motorway is 37.3 km in length, running from Nanjing to Lishui, with the starting point located near to Gaoqiaomen of the Nanjing Ring Road. The ending point connects with Guizhuang Hub, the starting point of Phase I of the Nanjing-Hangzhou Motorway.

11.2 Design principles

Based on analysis and understanding of the characteristics of the design objects, natural environment and cultural background the style of the green landscape design of a motorway is determined from a macro perspective, in order to ensure coordination between the motorway landscape and the natural landscape. Prior to design works, investigations have been carried out on the green landscapes of motorways opened to traffic in recent years. Based on a site investigation of Phase II of the Nanjing-Hangzhou Motorway and a comprehensive analysis of the environment along the route, the following design principles have been determined in combination with

The Environment and Landscape in Motorway Design, First Edition.
Qian Guochao, Tang Shuyu, Zhao Min and Jing Chun.
© 2014 China Communications Press. Published 2014 by John Wiley & Sons, Ltd.

natural conditions, the human environment, biological resources and climatic conditions.

People-oriented principle

Some sections of this Motorway run through cities and next to villages, so a people-oriented principle should always be implemented. For the landscape planning of motorways, driving safety requirements should be first met, and landscape design should then be used to improve the visual environment, relieve driver fatigue, create a comfortable driving environment and improve safety standards. In addition, attention should be paid to protecting the surrounding environment and giving full consideration to the feelings of drivers and passengers as well as surrounding residents, in order to achieve the harmonious integration of people, the motorway and the environment.

Principle of diversity and unity

As a part of the Nanjing-Hangzhou Motorway, Phase II of the Motorway is the main section from Nanjing to Hangzhou. Its starting point is an extension of the urban style, gradually integrating into the green landscape style of Phase I and maintaining the design philosophy of the 'pearl necklace'. In addition, the landscape design of different places features its own characteristics and reflects different geographical features, for example, Gaoqiaomen Hub and Shangfang Interchange reflect Qinhuai waterfront scenery; Dongshan Hub reflects the distinctive academic flavour of the Jiangning District Campus City; Hushu Interchange reflects bridges, streams and pavilions in spring; Guozhuang Interchange reflects Taoist culture; and Lishui Northern Interchange reflects a natural forest that blends into the surrounding mountainous forest landscape. Each node has its own characteristics and jointly presents an image of the historical and cultural city.

Principle of sustainable development

The green design of motorways should adapt to the natural terrain and topography in order to lower the amount of land levelling and reduce the impact of surface runoff and catchment areas caused by construction. Based on the concept of ecological restoration and ecological improvement, and combined with local ecological and environmental characteristics, the following works are carried out: protect and optimize the formed ecological areas and restore and reconstruct the damaged ecological areas as far as possible in order to achieve sustainable development. Since green landscapes have a strong dynamic nature, landscape design should consider certain continuity in the light of development. For plant selection, measures should be adjusted to local conditions, achieving a reasonable combination of fast-growing species and slow-growing species. Both the current effects and long-term effects should be taken into account, and a scientific development concept should be adopted to guide landscape design.

11.3 The details of green landscape design of the Nanjing-Hangzhou motorway (phase II)

Green landscape design of the central reservation

The main purposes of the central reservation are to divide lanes according to the direction of traffic, prevent headlight glare from disturbing drivers, relieve drivers' psychological sense of danger when passing by vehicles in the other direction, and reduce drivers' mental fatigue. In addition, central reservations can guide drivers' line of sight and improve the landscape. Central reservations should generally be designed to primarily feature uniform and integrated evergreen shrubs combined with the freestyle design of deciduous flowering shrubs. The ground surface is generally covered by flowers, ground cover plants and grass; the variation of different standard segments can relieve drivers' visual fatigue and passengers' sense of monotony.

Plant selection for the central reservation

Plants suitable for the central reservation should be able to absorb exhaust emissions and should be anti-pollution, robust-growing, slow-growing and resistant to trimming.

Anti-glare species:

Coniferous pagoda types: *Sabina komarovii* (large amount), *Juniperus chinensis* (small amount).
Broadleaf upright types: *Viburnum odoratissimum*, *Osmanthus fragrans* (small amount).
Spherical: *Photinia serrulata* (small amount).

Small trees:

Colourful leaf: Atropurpureum, *Prunus ceraifera*, *Acer palmatum*, etc.
Flowering shrubs: *Hibiscus syriacus*, *Malus micromalus* Makino, *Lagerstroemia indica*, *Punica granatum*, etc.

Ball (spherical) plants:

Green ball: *Pittosporum tobi*ra, *Buxus sinica*, *Ilex cornuta*, etc.
Red ball: *Loropetalum chinense* var. *rubrum*, etc.
Yellow ball: *Ligustrum ovalifolium* Vicaryi, etc.

Ground cover plants:

Open type: *Zephyranthes canndida*, *Ophiopogon japonicus* and grass dotted with flowers.
Closed type: *Photinia serrulata*, *Ligustrum ovalifolium* Vicaryi, *Rosa cultivars*, *Yucca gloriosa*, etc.

Arrangement option

See Figures 11.1–11.5 for arrangement options 1–5.

Arrangement option 1 (Figure 11.1): the strengthened sections are mainly arranged in service areas and urban sections. *Photinia serrulata* is adopted for glare prevention, which is planted in the centre at intervals of 5 m, with three plants of *Lagerstroemia indica* (which blooms in summer and autumn) and *Hibiscus syriacus* (which blooms in autumn) in between the balls. *Sabina chinensis* is planted at the outer edge of the guardrail, and the ground surface is fully covered by *Cynodon dactylon* in order to achieve a better seasonal landscape. For the island head section, *Photinia serrulata* is planted over a large area, and three rows of *Sabina chinensis* are planted at both sides in a staggered manner.

Arrangement option 2 (Figure 11.2): *Juniperus chinensis* and *Viburnum odoratissimum* or *Osmanthus fragrans* are used as the main anti-glare tree species. Three plants of *Juniperus chinensis* are jointly planted and look like a column, with a large canopy to increase the anti-glare effect. *Pittosporum tobira* and Atropurpureum are planted in a staggered way at both sides of the guardrail, which are both evergreen and change according to the season. *Canna indica* is planted on both sides of *Juniperus chinensis*, with *Photinia serrulata* planted at the edge to enrich the seasonal landscape. In addition, 30 m of *Viburnum odoratissimum* or *Osmanthus fragrans* is planted

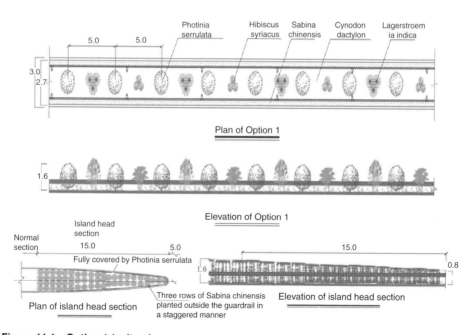

Figure 11.1 Option 1 (unit: m).

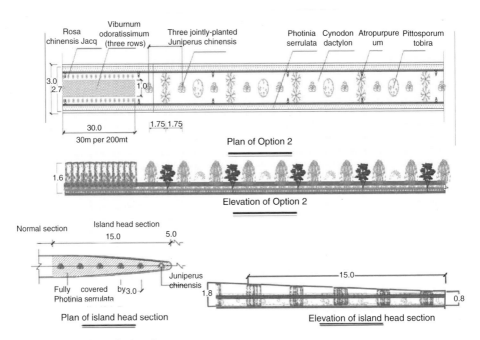

Figure 11.2   Option 2 (unit: m).

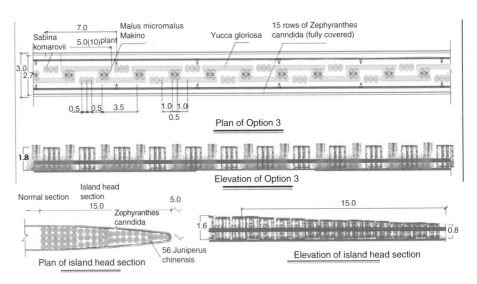

Figure 11.3   Option 3 (unit: m).

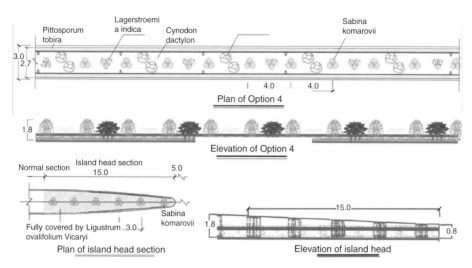

Figure 11.4 Option 4 (unit: m).

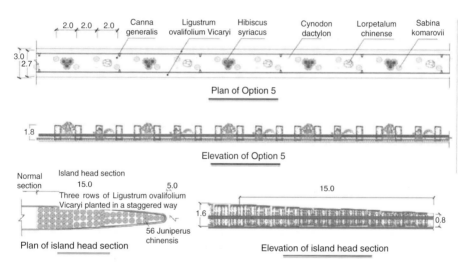

Figure 11.5 Option 5 (unit: m).

at intervals of 200 m. The surface is fully planted with *Cynodon dactylon* in order to achieve a refreshing and open landscape effect. For the island head section, 180–120 cm of *Juniperus chinensis* is planted and *Photinia serrulata* covers the surface.

Arrangement option 3 (Figure 11.3): three densely planted *Sabina komarovii* are used for glare prevention, which are planted in a staggered manner at both sides. *Yucca gloriosa* and *Malus micromalus* Makino are planted among *Sabina komarovii*, with *Yucca gloriosa* and *Sabina komarovii* forming a continuous high–low change to enrich the landscape in summer and autumn. *Malus micromalus* Makino in the centre can enrich the upper landscape in spring, with the ground surface fully covered by *Zephyranthes canndida*. For the island head section, *Juniperus chinensis* is trimmed into a slope-shape, and the ground surface is covered by *Zephyranthes canndida*.

Arrangement option 4 (Figure 11.4): three densely planted *Sabina komarovii* and two diagonally arranged *Ligustrum ovalifolium* Vicaryi form two dynamic layers. Three jointly planted *Lagerstroemia indica* are planted at intervals of 8 m to strengthen the landscape effect by their red decoration. *Pittosporum tobira* is used for decorating the edge of the outer side of the guardrail, fully covered with *Cynodon dactylon*. For the island head section, three jointly planted *Sabina komarovii* at a height of 180–120 cm are planted, fully covered by *Ligustrum ovalifolium* Vicaryi (80 cm).

Arrangement option 5 (Figure 11.5): two rows of *Sabina komarovii* are planted in a staggered manner, with a row of *Loropetalum chinense* var. *rubrum* (120 cm × 120 cm) planted in the middle. Three *Hibiscus syriacus* are planted instead of *Loropetalum chinense* var. *rubrum* at intervals of 20 m. Among the *Hibiscus syriacus*, *Canna generalis* is planted at intervals of 10 m. *Ligustrum ovalifolium* Vicaryi is planted at the outer edge of the guardrail, and the ground surface is fully covered by *Cynodon dactylon*. For the island head section, *Juniperus chinensis* is trimmed into a slope-shape, and three rows of *Ligustrum ovalifolium* Vicaryi are planted at the edge in a staggered manner.

The central reservation of the Nanjing-Hangzhou Motorway (Phase II) features grass dotted with flowers in order to achieve an evergreen effect and multi-coloured landscape, and to form a neat and orderly arrangement with a dynamic and rhythmic landscape effect (Figure 11.6).

Slope green landscape design

Slope landscape design should consider slope gradient, slope length, slope position and soil conditions, and combine with the surrounding natural and social environments. In addition to achieving landscape aesthetic effects, the design should integrate with engineering protection to ensure slope stabilization and avoid erosion.

Subgrade slope

Green landscape design of the subgrade slope: The following six arrangement options are mainly adopted for the greening of the subgrade slope.

Magnolia grandiflora+Prunus ceraifera+Jasminum nudirlorum *(height of fill section ≤2 m):* *Ginkgo biloba* and *Magnolia grandiflora* are planted at

Figure 11.6   **(a–f) Greening of central reservation.**

the outer edge of the slope and taken as the background. *Prunus ceraifera* and *Jasminum nudirlorum* are planted along the edge of the motorway, creating a plant landscape of flowers in spring and colourful leaves in autumn.

Nerium oleander+Hibiscus syriacus+Forsythia suspense *(height of fill section>2 m):*   For slopes of fill sections with a height greater than 2 m, *Salix babylonica* is planted at the outside of the slope, and *Sapium sebiferum* or

Magnolia grandiflora is planted at the inside of the slope. *Lagerstroemia indica*, *Hibiscus syriacus* or *Nerium oleander*, which bloom in summer and autumn, as well as *Cerasus serrulata* or *Forsythia suspense*, which bloom in spring, are planted along the edge of the motorway.

Sapium sebiferum+Cerasus serrulata+Lagerstroemia indica+*grass dotted with flowers (height of fill section >2 m):* People can enjoy the exotic diamond leaves of *Sapium sebiferum* and its red leaves in autumn. The natural planting of *Cerasus serrulata* (which blooms in spring) and *Lagerstroemia indica* (which blooms in summer) as well as grass dotted with colourful flowers presents a varying seasonal landscape.

Ginkgo biloba+Zelkova schneideriana+Prunus ceraifera+Osmanthus fragrans *cv. Semperflo*+Zephyranthes canndida*:* Through natural-style planting methods, trees, shrubs and ground cover plants as well as flowers, leaves and fragrant plants are organically combined in order to fully reproduce the natural landscape.

Ailanthus altissima+Hibiscus syriacus+*Atropurpureum*+Photinia serrulata *(height of fill section >2 m):* Taking *Ailanthus altissima* as the background, Atropurpureum, *Photinia serrulata* and *Hibiscus syriacus* (which blooms in summer) are planted.

Cedrus deodara+Prunus persica+Hibiscus syriacus *(height of fill section >2 m):* Taking beautiful evergreen *Cedrus deodara* as the background, *Prunus persica*, which blooms in spring, and *Hibiscus syriacus*, which blooms in summer are planted, presenting a scene of vitality.

Green landscape effect of the subgrade slope: The greening effect of the subgrade slope is shown in Figure 11.7.

Figure 11.7   (a, b) Greening effect of subgrade slope.

Cutting slope

Ecological protection of the cutting slope lies in slope stabilization and improving the anti-erosion capacity of the slope surface, which can conserve water and reduce soil erosion and can also purify the air and protect the ecological environment, achieving good economic, social and ecological benefits.

Green landscape design scheme of the cutting slope: The following five options are mainly given for the greening of cutting slopes on the Nanjing-Hangzhou Motorway (Phase II).

Zelkova schneideriana+Cinnamomum camphora+*coloured-leaf trees*+ *flowering shrubs*: Taking beautiful deciduous trees and evergreen trees as the background, coloured-leaf trees such as Atropurpureum, *Helicia cochinchinensis* and *Photinia serrulata* are planted at the upper part, central part and lower part of the slope, with *Malus halliana* blooming in spring jointly planted in order to create a natural-style landscape with both colourful leaves and flowers for people to enjoy.

Bamboo+Pinus elliottii: According to existing vegetation, pine and bamboo are mainly planted. *Phyllostachys nidularia* and *Pinus elliottii* can form a natural landscape, decorated by *Forsythia viridissima* and grass dotted with flowers, reflecting the natural and wild landscape.

Phyllostachys nidularia+Canna generalis: The natural and clump planting of evergreen bamboo combined with *Canna generalis* enriches the layering of the plants by the natural integration of red and green colours.

Bamboo+Yucca gloriosa+*coloured-leaf trees*: The well-growing *Phyllostachys nidularia* is planted to form a forest belt, and combined with *Yucca gloriosa* and coloured-leaf trees such as Atropurpureum to establish natural and stable plant communities. This is achieved by a combination of tall and short and near and distant plants. In autumn the fiery-red scene accentuates the natural landscape.

Landscaping trees+grass: *Zelkova schneideriana* and *Albizia julibrissin* are planted in clumps, which form groups every 100 m, contrasting very well with each other. The mixed planting of *Lolium perenne*, *Zoysia Matrella* and *Trifolium repens* at the lower section creates an open woodland and grassland landscape.

Green landscape effect of the cutting slope: The greening effect of the cutting slope is shown in Figure 11.8.

(a)

(b)

(c)

(d)

(e)

(f)

Figure 11.8   (a–f) Greening effect of the cutting slope.

Green landscape design of overpass bridges

Overpass bridges are one of the areas of motorways which have an important visual impact on drivers and passengers. The vertical greening method is adopted at the piers of landscaping bridges in Phase II of the Nanjing-Hangzhou Motorway in order to enrich the bridge landscape.

Plant arrangement of the Aidong Highway Grade Separation without ramps (Figure 11.9)

The mass-planting of the large tree *Ginkgo biloba* forms a special landscape, and original vegetation is selected as the background forest in order to ensure integration with the surrounding environment. They are combined with *Cedrus deodara*, *Pterocarya stenoptera*, *Magnolia grandiflora*, *Ligustrum lucidum* Aiton, *Firmiana simplex*, *Prunus persica*, *Prunus lannesiana*, *Hibiscus syriacus*, *Photinia serrulata*, *Jasminum mesnyi* and *Rosa multiflora*. This kind of natural planting creates an evergreen landscape with a plant community that is in bloom for three seasons of the year. A vertical greening method is adopted at the piers of the landscaping bridges, and *Parthenosissus tricuspidata* and Lardizabalaceae are planted to improve the appearance of the cement structures.

Plant arrangement of Lishui Development Zone Grade Separation (Figure 11.10)

The mass planting of *Pinus elliottii* forms a special landscape. Based on the original vegetation, *Elaeocarpus decipiens*, *Koelreuteria paniculata*, *Chimonanthus praecox*, *Forsythia viridissima* and *Phyllostachys nidularia* as well as grass dotted with flowers are planted in order to form plant communities combining trees, shrubs and grasses and integrate with the surrounding environment. Mesh suspended and vertical greening is adopted at the piers of the landscaping bridge, and *Hedera nepalensis* and *Trachlospermum jasminoides* are planted to enhance the bridge landscape.

Figure 11.9   Greening effect of the Aidong Highway Grade Separation without ramps.

Figure 11.10   Greening effect of at the Shenshui Development Zone Grade Separation.

Plant arrangement of Zhannan Road Grade Separation (Figure 11.11)

The mass planting of *Prunus mume* forms a special landscape, and based on the original vegetation, *Koelreuteria paniculata, Osmanthus fragrans*, Atropurpureum, ground cover plants, grass, flowers, *Gaillardia aristata* and *Zephyranthes canndida* are arranged to form plant communities that combine trees, shrubs and grasses and integrate with the surrounding environment. The landscaping trees beside the road have a framing effect, forming a unique landscape.

Figure 11.11   Greening effect of Zhannan Road Grade Separation.

Green landscape design of interchange hubs

The interchange hubs of the Nanjing-Hangzhou Motorway (Phase II) follow the design philosophy of the 'pearl necklace' in Phase I, regarding each interchange as a 'pearl' embedded in the 'long necklace' motorway. Tailored to the characteristics of each section, a distinctive natural landscape is created. After a certain amount of treatment, the original natural landscape is incorporated into the visual range of drivers and passengers. This involves using special techniques of extending space in order to arrange and expand space, and using the original pond slope within the interchange for topographical improvement, before supplementing with plants to form a landscape that reflects the local environment. For the green landscape arrangement of interchange areas, full consideration should be given to coordination with the surrounding environment, making sure that there is a free and orderly arrangement of a variety of tall and short plants. Each interchange area has its own unique landscape features, so drivers and passengers can enjoy the pleasant plant landscape and feel the overall effect of the dynamic landscape. Some shrubs can be planted at the outside of the ramp for guiding drivers' vision.

Design ideas

According to the current situation and characteristics of the surrounding environment, each interchange hub should be designed with distinctive features and a different theme:

- Gaoqiaomen: graceful Qinhuai style along a 'long-bead curtain' zone – dominated by a summer and autumn landscape.
- Shangfang: wetland landscape – dominated by a spring and autumn landscape.
- Dongshan: academic flavour of Jiangning District Campus City – dominated by an autumn landscape.
- Hushu: warm river water in spring and picturesque islands – dominated by a spring and summer landscape.
- Guozhuang: Taoist culture – dominated by a summer and autumn landscape.
- Northern Lishui: natural forest – dominated by a summer and autumn landscape.

Plant arrangement modes

Gaoqiaomen hub: The green landscape design of Gaoqiaomen Hub follows a design idea that focuses on natural landscapes. This forms the overall layout that corresponds to scenery on both sides of the road, reflecting the design theme of 'Graceful Qinhuai style along a "long-bead curtain" zone'. At the junction of land and water, a natural waterfront is arranged, and aquatic plants are planted in order to create a changing wetland landscape. The west side of the ring road mainly features a waterfront landscape, with wetland plants including *Pinus elliottii*, *Salix babylonica* and *Hibiscus mutabili*s as well as aquatic plants including *Nymphaea alba*, *Phragmites australis*, *Iris pseudacorus* and *Thalia dealbata*. A few rocks are used to decorate the landscape; on the east side, there is a microtopographic slope, on which *Cedrus deodara* and *Osmanthus fragrans* are planted at the higher layer, combined with *Ginkgo biloba* and *Zelkova schneideriana* and inter-planted with the flowering shrubs of *Prunus persica* 'Atropurpurea' and *Lagerstroemia indica*, and *Zinnia eIegans* and *Coreopsis basalis*, which bloom in summer and autumn, are planted at the lower layer. Communities that feature a combination of trees, shrubs and grasses as well as woodland and grassland are formed at both sides of the commanding height. In general, a summer plant landscape should dominate along sections leading into cities, planting species such as *Lagerstroemia indica*, *Hibiscus syriacus* and *Albizia julibrissin* to create a warm, lively and welcoming atmosphere. Autumn fruit plants should dominate along sections leading out of cities, planting species such as *Koelreuteria paniculata*, *Punica granatum* and *Pyracantha fortuneana*, to reflect the autumn harvest.

Shangfang interchange: The interchange design area is traversed by a river, surrounded by natural ponds and features a generally low-lying terrain.

The landscape design of Shangfang Interchange is focused on creating a natural waterbody landscape and ensuring unity and coordination with the environment. Approaching the overpass of Dongxian Road, mixed forest composed of tall *Ginkgo biloba* and *Cinnamomum camphora* is used as the background forest, and the flowering shrubs *Prunus mume* and *Malus spectabilis* are planted at the edge of the forest, in order to highlight the characteristics of Nanjing city. The magnificent bridge contrasts well with the landscape, which enriches the overall landscape effect of the interchange.

Around the main line, a large area of trees such as *Ginkgo biloba* and *Cinnamomum camphora* as well as Atropurpureum, *Photinia serrulata* and *Osmanthus fragrans* cv. Semperflo forms a view point. At the waterfront close to the main line, the hydrophilous plants *Salix babylonica* and *Sapium sebiferum* together with ground cover plants and flowering shrubs form an elegant landscape with weeping *Salix babylonica* on the banks. Additionally, hydrophilous flowers and aquatic flowers reflect the unique natural landscape of the Yangtze River Delta.

Dongshan hub: Dongshan Hub is located among hills and mountains, at the intersection of the Nanjing-Hangzhou Motorway and Nanjing Ring Road; Jiangning District Campus City is in its vicinity.

Dongshan Hub is designed to reflect a scholarly atmosphere in both its layout and elevation. The water surface across this area and the mass-planting of

Nandina domestica form a 'heart' shape, implying that diligence is required when studying, working and communicating with others. From the elevation, the plant landscape and its poetic connotations intensify the academic atmosphere. Two large areas of water at both sides of the main line form the core landscape of this hub, containing a landscape that features terrestrial, wetland and aquatic plants (Figure 11.12).

At the area enclosed by Ramp A with the main line: *Cinnamomum camphora*, *Koelreuteria paniculata*, *Sapium sebiferum* and *Pterocarya stenoptera* form the background forest. *Metasequoia glyptostroboides*, *Albizia julibrissin* and *Jasminum mesnyi* as well as the aquatic flowers *Thalia dealbata*, *Iris tectorum*, *Arundo donax* var. *versicolor*, *Phragmites australis* and *Scirpus tabernaemontani* planted at the waterfront form a plant community. At the relative landside, *Cedrus deodara*, *Sapium sebiferum*, *Magnolia liliflora*, *Nerium oleander*, *Pyracantha fortuneana*, *Loropetalum chinense* var. *rubrum* and *Rhododendron pulchrum* form different terrestrial plant communities.

At the triangular area enclosed by Ramp B with the main line: *Catalpa bungei*, *Pterocarya stenoptera*, *Salix babylonica* and *Malus halliana* are planted according to the terrain, creating a landscape of woodland and grassland dominated by a spring and summer plant landscape, making for a welcoming atmosphere for visitors.

At the area enclosed by Ramp C and the main line: *Cinnamomum camphora*, *Catalpa bungei*, *Ascendens mucronatum*, *Robinia hisqida*, *Prunus*

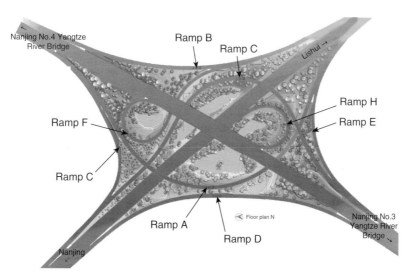

Figure 11.12   Plan of Dongshan Hub.

persica, Atropurpureum and *Kerria japonica* are planted at one side of the waterfront, forming an attractive community landscape all year round combined with the aquatic flowers *Nymphaea alba*, *Miscanthus*, *Iris pseudacorus*, *Sagittaria sagittifolia*, *Arundo donaxl* and *Lythrum salicaria*. At the landside, with the broadleaf trees *Cinnamomum camphora* and *Catalpa bungei* as the background, *Juniperus chinensis*, *Prunus* × *cistena*, *Mahonia fortune*, *Althaea rosea*, *Cosmos bipinnatus* and *Cosmos sulphureus* are planted, forming a landscape different from the terrestrial community landscape between the main line and Ramp A.

At the triangular area and other areas enclosed by Ramp D, Ramp E, Ramp B, Ramp G and the Ring Road: the seasonal plant landscape of the water within the central part is extended, and a grouped landscape of woodland and grassland is created according to the terrain. *Liriodendron chinense*, *Rhus typhina* Nutt, *Photinia serrulata* and *Yucca gloriosa* are planted. Additionally, at the triangular area enclosed by Ramp D, Ramp G and the Ring Road, pine, bamboo and plum are planted to enhance the attractive landscape.

Hushu interchange: Due to abundant water within the Hushu Interchange and in order to highlight the features of a rural water landscape, aquatic plants are planted in areas of shallow water and at the waterfront, creating a wetland landscape with *Hibiscus mutabilis*, *Phragmites australis*, *Iris pseudacorus* and *Thalia dealbata*. Undulating terrain is conducive to plant layout and landscaping, and trees can be naturally planted to form a community, such as *Cedrus deodara*, *Cinnamomum camphora*, *Zelkova schneideriana*, *Elaeocarpus decipiens* and *Osmanthus fragrans* as well as *Lagerstroemia indica*, Atropurpureum, *Nerium oleander* and *Nandina domestica*. These plants have attractive flowers and leaves and make the whole interchange seem peaceful and natural from afar. Grass dotted with flowers is arranged around the interchange, turning the landscape into a sea of flowers. The green space of the Hushu Interchange has multiple layers, resulting in a strong three-dimensional feeling. There is a combination of trees, shrubs and grasses, creating an evergreen landscape with flowers that bloom in spring, summer and autumn. Furthermore, the integration of terrestrial plants and wet plants as well as the arrangement of evergreen plants, flowering plants, fruiting plants and ornamental foliage plants makes each area and each view within the interchange in harmony.

Guozhuang interchange: A wetland landscape with local characteristics can be created by taking full advantage of local native species. To cope with the effects of topography, the densely planted trees at the extruded large area in the central part are taken as the background, with the broadleaf plants *Cinnamomum camphora* and *Sapium sebiferum* used as the dominant tree species. *Salix babylonica*, *Liquidambar formosana* and *Hibiscus mutabilis* are planted along the water surface, creating a hierarchical and ecological plant landscape with trees and shrubs such as *Prunus ceraifera* and *Pyracantha* *fortuneana*. This mixed forest is then extended to the two triangular areas to form a single mixed forest. *Metasequoia glyptostroboides* forest dominates at the small circular area of the interchange. Flowering trees and shrubs such as *Osmanthus fragrans* cv. Semperflo, *Malus halliana*, *Nerium oleander*, *Hibiscus mutabilis* and *Prunus ceraifera* are mass-planted at the edge of the forest in the interchange, which corresponds to the surrounding natural rural scenery and also creates a splendid landscape with spectacular flowers in summer and autumn. Furthermore, the group planting of short shrubs and ground cover plants below the forest and at the edge form a natural community landscape within the interchange.

North Lishui interchange: From the Nanjing–Lishui direction along the main line, a *Prunus persica* (var. densa Makino) forest can be seen, slowly guiding people's sight to the water ahead. A mixed forest of *Cinnamomum camphora*, *Sapium sebiferum* and *Koelreuteria integrifoliola* forms the narrow background woodland of the two long triangular areas of water. A group of *Phyllostachys nidularia* is planted at the head of the bridge to enrich the understory landscape. The landscape on both sides of the bridge appears coordinated and in unity. The triangular area close to the large area is the confluence point of the sight, where sparsely planted trees on the grassland form a landscape of woodland and grassland. *Cedrus deodara* woods on one side of the bend of the large area can be seen through the woodland and grassland. A stretch of *Albizia julibrissin* woods is close to the waterside at the back of the *Cedrus deodara* forest, with abundant red flowers that bloom in midsummer at the waterfront, creating a good visual effect when seen from the main line and the toll station.

At the elongated island in the large area, the plant arrangement follows that of a mountain landscape, with mixed forest dominated by *Cinnamomum camphora*, *Celtis sinensis*, *Ginkgo biloba* and *Koelreuteria integrifoliola* forming the background of the whole interchange. Two patches of *Nelumbo nucifera* are planted at the water surface, which casts beautiful reflections with *Albizia julibrissin* at the waterfront in summer. Two points are provided at the main line and the ramp for enjoying the scenery.

From the Lishui–Nanjing direction along the main line, a stretch of *Pinus elliottii* woods is the first sight that is seen, which echoes with the mountain forest outside; opposite there is a mixed forest of *Celtis sinensis* and *Cinnamomum camphora*, which integrates into the surrounding environment. A large area of water is located beside the main line, where *Lythrum salicaria* is planted. A landscape of woodland and grassland is created at the triangular area close to the circular area, with an isolated tree planted to echo with the isolated trees of the main line opposite, meaning that the views at both sides of the main line complement each other well.

Forest and valleys form the dominant view at the circular area. At the lowland between the two high points there are scattered rocks and tree forest at both sides. There is mixed forest of *Celtis sinensis* and *Cinnamomum camphora* at one side close to the bridge. Mixed woods of *Cinnamomum camphora* and *Koelreuteria paniculata* are formed at the central high point, and water-tolerant *Ascendens mucronatum* is planted in the lowland area, forming rich canopy lines.

For the design of the overall interchange, mixed forest is adopted instead of natural mountain forest in order to form a harmonious landscape with the surrounding mountains and woods.

Landscape effect of interchange hub

See Figures 11.13–11.20 for the landscape effect of the interchange hub.

Green landscape design at service areas

Service areas provide a resting place for motorway users as well as refreshment and vehicle maintenance services, and their main function is to meet the leisure demands of travellers in order to relieve fatigue. For this reason, their green design should focus on static landscape design as well as a combination

(a)

(b)

(c)

(d)

Figure 11.13   **(a–d) Local green landscape of Gaoqiaomen Hub.**

Figure 11.14   Green landscape effect of Shangfang Interchange.

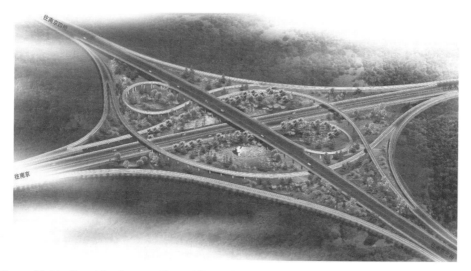

Figure 11.15   Green landscape effect of Dongshan Hub.

of static and dynamic landscapes, using the concept of scientific development to guide green landscape design. The greening of service areas should take the functional requirements of each part into consideration, and adapt to local conditions. For instance, tall trees should be planted in parking areas to provide shade for vehicles. For buildings and squares at service areas,

(a)   (b)

(c)   (d)

Figure 11.16   (a–d) Local green landscape of Dongshan Hub.

garden-style methods can be used for constructing raised flower terraces in order to strengthen the aesthetic effect and create a comfortable resting envi ronment. In addition, for office zones and living zones, trees, shrubs, flowers and grass should be planted appropriately to create a pleasant environment.

Selection of tree species

Plants for service areas should be colourful and attractive, with strong resilience and conducive to planting, survival, trimming and management. A scientific arrangement of trees, shrubs, flowers and grasses should be considered to create a seasonal landscape featuring 'flowers in spring, shade in summer, harvest in autumn and greenery in winter'. The plants should be dominated by native species in order to reflect the characteristics of Nanjing.

Specific design forms

The overall layout and landscape design of service areas should focus on streamlined design with the parking areas located at one end of the service area. It is also important to ensure the effective separation of parking lots for heavy vehicles, light vehicles and trailers by green space, and integrating

(a)

(b)

(c)

(d)

Figure 11.17   (a–d) Local green landscape of Hushu Interchange.

Figure 11.18   Green landscape effect of Guozhuang Interchange.

Figure 11.19   Green landscape effect of Northern Lishui Interchange.

Figure 11.20   (a–d) Local green landscape of Northern Lishui Interchange.

the whole parking area into the densely planted trees. The plants around the parking area should be mainly *Zelkova schneideriana*, *Elaeocarpus decipiens*, *Osmanthus fragrans* cv.Semperflo and ground cover plants. Tree-lined pools, flower beds and flower columns are provided at the front of the main buildings, casting beautiful reflections with the main buildings and creating an attractive and comfortable service area environment together with natural plant arrangements. Due to the importance of the service area, there are strict landscape requirements. During plant landscaping, priority should be given to native species to better reflect local characteristics. The selected plants should have ornamental value and should be combined with trees, shrubs, flowers and grasses that feature significant seasonal changes. The colours of shoots, branches, flowers and leaves should be used for plant landscaping.

11.4 Conclusion

The construction of green landscapes on the Nanjing-Hangzhou Motorway (Phase II) adheres to the design philosophy of 'respect nature, protect nature and restore nature', learns from successful experiences and implements innovation. The overall green landscape design is tailored to nature, striving to integrate the motorway into the surrounding natural environment and become a part of the natural landscape. The landscape is formed according to the terrain conditions along the motorway, meaning there are no visual obstacles. No crash barriers are constructed at some sections so that motorway users can fully enjoy the artificial and natural scenes. The motorway seems to be embedded in the forest, grassland and wetland, and is both natural and attractive. In addition, according to the achievements in ecological protection during the construction of the Nanjing-Hangzhou Motorway Phase I, ordinary seeding, mesh-suspended spray seeding, hole-planting of pine and bamboo and other ecological protection methods have been adopted in Phase II in order to ensure soil consolidation and slope protection as well as greening and landscaping. Also, based on the use of grass dotted with flowers in Phase I, grass dotted with flowers has been planted at interchange hubs and slopes for Phase II, and used in a pilot application in central reservations, in order to enrich the colour of the motorway and create an appealing natural environment for drivers and passengers. In different seasons, the plant landscape reflects the harmonious combination of an artificial and natural landscape.

Based on plant growth patterns, characteristics and habits, suitable plants are selected and arrangement modes are optimized for the green landscape design of Phase II, not only providing a safe, comfortable, smooth, efficient and pleasant driving environment for motorway users but also adding a unique cultural element to the landscape of the motorway, improving the quality of the environment and ensuring harmony between the historical, cultural and natural landscape. The completion Phase II of the Nanjing-Hangzhou Motorway represents another new model in motorway construction in Jiangsu Province and in China as a whole, and reflects a perfect combination of humanity and ecology. It is the best motorway in Jiangsu Province, featuring the best in quality, greening effects and landscapes.

References

An, S.Q., Zhao, R.L. Zhong guo bei ya re dai ci sheng sen lin zhi bei de te zheng fen xi. *Nan jing da xue xue bao (zi ran ke xue ban)*, 1991, **2**

Cheng, J. *Yuan ye*. Chongqing: Chongqing Publications, 2009

He, J.S., Chen, W.L., Li, L.H. Zhong guo zhong ya re dai dong bu chang lü kuo ye lin zhu yao lei xing de qun luo duo yang xing te zheng. *Zhi wu sheng tai xue bao*, 1998, **4**

Lynch, K. *The Image of the City*. Cambrige, MA: MIT Press, 1960

Yu, S.H. *Zhe jiang sheng chang lü kuo ye lin de sheng tai xue yan jiu*. Beijing: Bei Jing lin ye da xue bo shi xue wei lun wen, 2003

Zhu L.M., Feng Z.W. Jiang su ma wei song lin xia mu ben zhi wu qun luo xiang si xing he sheng wu duo yang xing de diao cha. *Huan jing yu kai fa*, 1998, **1**

Further Reading

Baas, P., Kalkman, K., Geesink, R. (eds) *The Plant Diversity of Malesia*. Dordrecht: Kluwer Academic, 1990

Bi, J., Ma, Z.W., Xu, Y.L. deng. Ren gong ce bai qun luo jie gou ji sheng wu liang. *Shenyang: dong bei lin ye da xue xue bao*, 2000, **28**(1). 毕君, 马增旺, 许云龙, 等. 人工侧柏群落结构及生物量. 沈阳 : 东北林业大学学报, 2000, **28**(1)

Cai, X.H., Wang, J.X. Bu tong gan rao cheng du dui chang lü kuo ye lin ji di qun luo jie gou de ying xiang. *Si chuan lin ye ke ji*, 2003, **24**(1). 蔡小虎, 王金锡. 不同干扰程度对常绿阔叶林迹地群落结构的影响. 四川林业科技, 2003, **24**(1)

Cao, L.Y. Gao su gong lu jian she de sheng tai huan jing wen ti ji dui ce yan jiu. *Shan xi jiao tong ke ji*, 2002, **4**. 曹力媛. 高速公路建设的生态环境问题及对策研究. 山西交通科技, 2002, **4**

Chen, F., Yu, Q.C. Shi lun gong lu jiao tong fa zhan dui sheng tai huan jing de ying xiang ji dui ce. *Wu han jiao tong ke ji da xue xue bao (she hui ke xue ban)*, 2000, **13**(4), 10. 陈芬, 郁麒昌. 试论公路交通发展对生态环境的影响及对策. 武汉交通科技大学学报 (社会科学版), 2000, **13**(4), 10

Chen, H.W., Liu, Y.G., Feng, X. deng. Yun nan xi shuang ban na xi nan hua ren gong lin qun luo jie gou chu bu yan jiu. *Guang xi lin ye ke xue*, 1999, **28**(3). 陈宏伟, 刘永刚, 冯弦, 等. 云南西双版纳西南桦人工林群落结构初步研究. 广西林业科学, 1999, **28**(3)

Chen, Y.M. *Yuan lin shu mu xue*. Beijing: Zhong guo lin ye chu ban she, 1990. 陈有民. 园林树木学. 北京 : 中国林业出版社, 1990

Cheng, S.G. *Gao su gong lu huan jing ping jia yu fa zhan*. Beijing: Zhong guo huan jing ke xue chu ban she, 2002, **12**. 程胜高. 高速公路环境评价与发展. 北京 : 中国环境科学出版社, 2002, **12**

Cheng, Z.H., Zhang, J.T. Sheng tai lü you qu bu tong ju li dai shang zhi wu qun luo de jie gou dui bi. *Ying yong yu huan jing sheng wu xue bao*, 2002, **8**(1). 程占红, 张金屯. 生态旅游区不同距离带上植物群落的结构对比. 应用与环境生物学报, 2002, **8**(1)

Cui, W.B. *Gao su gong lu jing guan yan jiu chu tan*. Nanjing: Nan jing lin ye da xue shuo shi lun wen, 2003. 崔文波. 高速公路景观研究初探. 南京 : 南京林业大学硕士论文, 2003

Ding, S.Y., Song, Y.C. Zhe jiang tian tong guo jia sen lin gong yuan chang lü kuo ye lin yan ti qian qi de qun luo sheng tai xue te zheng. *Zhi wusheng tai xue bao*, 1999, **23**(2). 丁圣彦, 宋永昌. 浙江天童国家森林公园常绿阔叶林演替前期的群落生态学特征. 植物生态学报, 1999, **23**(2)

Ellenberg, H., Mueller-Dombois, D., Bao, X.C. deng. *Zhi bei sheng tai xue mu di he fang fa*. Beijing: Ke xue chu ban she, 1986. D. 米勒-唐布依斯, H. 埃仑伯格著, 鲍显诚, 等译. 植被生态学的目的和方法. 北京 : 科学出版社, 1986

Fan, F. *Qiao Liang mei xue*. Beijing: Ren min jiao tong chu ban she, 1987. 樊凡. 桥梁美学. 北京：人民交通出版社, 1987

Feng, W. *Xian dai jing guan she ji jiao cheng*. Hangzhou: Zhong guo mei shu xue yuan chu ban she, 2002. 冯炜. 现代景观设计教程. 杭州：中国美术学院出版社, 2002

Feng, Z.W., Wang, X.K., Wu, G. *Biomass and Net Primary Productivity of Forest Ecosystems in China*. Beijing: Science Press, 1999

Fu, B.J., Chen, L.X., Ma, K.M., Wang, Y.L. *Jing guan sheng tai xue yuan li ji ying yong*. Beijing: Ke xue chu ban she, 2001. 傅伯杰, 陈利顶, 马克明, 王仰麟. 景观生态学原理及应用. 北京：科学出版社, 2001

Gao, B.L. *Cheng shi gui dao jiao tong gao jia qiao de xuan xing she ji ji shi jian*. Beijing: Bei fang jiao tong da xue shuo shi lun wen, 2002. 高宝林. 城市轨道交通高架桥的选型设计及实践. 北京：北方交通大学硕士论文, 2002

Hao, Z.Q., Yu, D.Y., Wu, G. deng. Chang bai shan bei po zhi wu qun luo β duo yang xing fen xi. *Sheng tai xue bao*, 2001, **21**(12). 郝占庆, 于德永, 吴钢, 等. 长白山北坡植物群落β多样性分析. 生态学报, 2001, **21**(12)

He, P.Z. *Qiao Liang mei xue*. Beijing: Ren min jiao tong chu ban she, 1999. 和丕壮. 桥梁美学. 北京：人民交通出版社, 1999

He, Z.Y., Zhang, X.N., Shi, W.Z. Jing guan sheng tai xue zai gong lu jing guan huan jing ping jia zhong de ying yong. *Chng shi huan jing yu cheng shi sheng tai*, 2003, **16**(6), 133–135. 贺志勇, 张肖宁, 史文中. 景观生态学在公路景观环境评价中的应用. 城市环境与城市生态, 2003, **16**(6), 133–135

Hu, D.Q. Guang zhou shi jin jiao nong tian tu rang ji zuo wu qian wu ran shui ping ping jia. *Sheng tai ke xue*, 1997, **16**(1), 71–74. 胡迪琴. 广州市近郊农田土壤及作物铅污染水平评价. 生态科学, 1997, **16**(1), 71–74

Hu, Z.H., Yu, M.J., Fang, T. Zhe jiang gu tian shan zi ran bao hu qu chang lü kuo ye lin qun luo te zheng. *Nan jing qi xiang xue yuan xue bao*, 2003, **26**(1). 胡正华, 于明坚, 方腾. 浙江古田山自然保护区常绿阔叶林群落特征. 南京气象学院学报, 2003, **26**(1)

Huang, B.L. *Jiang su sen lin*. Nanjing: Jiang su ke xue ji shu chu ban she, 1998 黄宝龙. 江苏森林. 南京：江苏科学技术出版社, 1998

Huang, J.H., Li, Q., Liu, X.L. He nan zhou kou zhi sheng jie duan gao su gong lu jian she dui sheng tai huan jing de ying xiang. *Sheng tai xue za zhi*, 2002, **21**(1), 174–179. 黄锦辉, 等. 河南周口至省界段高速公路建设对生态环境的影响. 生态学杂志, 2002, **21**(1), 174–179

Hughes, R.G. Theories and models of species abundance. *The American Naturalist*, 1986, **128**, 879–899

Hung, H., Xie, Y., Huang, Z.X. Jiang xi jing gang shan tian zhu qun luo te zheng ji wu zhong duo yang xing. *Huan jing yu kai fa*, 2001, **16**(1). 黄虹, 谢宇, 黄兆祥. 江西井冈山甜槠群落特征及物种多样性. 环境与开发, 2001, **16**(1)

Jiang, Y.X., Wang, B.S., Zang, R.G. deng. *Hai nan dao re dai lin sheng wu duo yang xing ji qi xing cheng ji zhi*. Beijing: Ke xue chu ban she, 2002. 蒋有绪, 王伯荪, 臧润国, 等. 海南岛热带林生物多样性及其形成机制. 北京：科学出版社, 2002

Jin, J. *Li xiang jing guan-cheng shi jing guan kong jian xi tong jian gou yu zheng he sheji*. Nanjing; Dong nan da xue chu ban she, 2003. 金俊. 理想景观－城市景观空间系统建构与整合设计. 南京：东南大学出版社, 2003

Jin, L.F., Liu, X.P. 3S ji shu zai feng jing qu gui hua zhong de ying yong yan jiu. *Zhong guo yuan lin*, 1997, **13**(6). 金丽芳. 刘雪萍. 3S技术在风景区规划中的应用研究.中国园林, 1997, **13**(6)

Kempton, R.A. The structure of species abundance and measurement of diversity. *Biometrics*, 1979, **35**, 307–321

Lan, S.R. Fu zhou guo jia sen lin gong yuan ren gong qun luo jie gou yu wu zhong duo yang xing. *Fuzhou: fu jian lin xue yuan xue bao*, 2002, **22**(1), 1–3. 兰思仁. 福州国家森林公园人工群落结构与物种多样性.福州 : 福建林学院学报, 2002, **22**(1), 1–3

Leonhardt, F. *Qiao liang jian zhu yi shu yu zao xing*. Beijing: Ren min jiao tong chu ban she, 1988. [德]弗里茨·莱昂哈特. 桥梁建筑艺术与造型. 北京 : 人民交通出版社, 1988

Li, J.Y. Lun gao su gong lu lü hua li di tiao jian de te yi xing. *He bei lin ye ke ji*, 2000, **1**. 李俊英. 论高速公路绿化立地条件的特异性. 河北林业科技, 2000, **1**

Li, J.Y. Lun gao su gong lu lü hua li di tiao jian de te yi xing. *He bei lin ye ke ji*, 2000, **1**. 李俊英. 论高速公路绿化立地条件的特异性. 河北林业科技, 2000, **1**

Li, M.L., Wu, Z.Z., Ma, Y.F. Tu gong ge zha zai dao lu bian po fang hu zhong de ying yong. *Liao ning jiao tong ke ji*, 2001, **24**(5). 李美玲, 等. 土工格栅在道路边坡防护中的应用. 辽宁交通科技, 2001, **24**(5)

Li, R., Li, B.C. Gao su gong lu lü hua she ji chu tan. *Gong lu*, 1997, **7**. 李睿炬. 李斌成. 高速公路绿化设计初探. 公路, 1997, **7**

Li, W.Y., Xiang, Y.H., Du, Y. Yun nan zhan yi hai feng zi ran bao hu qu di xia sen lin qun luo te zheng fen xi. *Xi nan lin xue yuan xue bao*, 2001, **21**(1). 李伟云, 向艳辉, 杜宇. 云南沾益海峰自然保护区地下森林群落特征分析. 西南林学院学报, 2001, **21**(1)

Li, X.D., Zhai, N., Guang, C. San wei zhi bei gu tu wang dian zai lu ji bian po fang hu shang de ying yong fen xi. *Dong bei gong lu*, 2000, **23**(3). 李旭东, 等. 三维植被固土网垫在路基边坡防护上的应用分析. 东北公路, 2000, **23**(3)

Li, Z., Guo, S.Y. Ren wei shui tu liu shi yin su ji fang zhi cuo shi yan jiu. *Shui tu bao chi tong bao*, 1998, **18**(2), 48–52. 李智广, 郭索彦. 人为水土流失因素及其防治措施研究. 水土保持通报, 1998, **18**(2), 48–52

Liu, C.R., Ma, K.P., Lü, Y.H. Sheng wu qun luo duo yang xing de ce du fang fa VI-Yu duo yang xing ce du you guan de tong ji wen ti. *Sheng wu duo yang xing*, 1998, **6**(3), 229–239. 刘灿然, 马克平, 吕延华等. 生物群落多样性的测度方法VI-与多样性测度有关的统计问题. 生物多样性, 1998, **6**(3), 229–239

Liu, C.R., Ma, K.P. Sheng wu qun luo duo yang xing de ce du fang fa V-Sheng wu qun luo wu zhong shu mu de gu ji fang fa. *Sheng tai xue bao*, 1997, **17**(6). 刘灿然, 马克平. 生物群落多样性的测度方法 V-生物群落物种数目的估计方法. 生态学报, 1997, **17**(6)

Liu, C.R., Ma, K.P., Yu, S.L. deng. Bei jing dong ling shan di qu zhi wu qun luo duo yang xing de yan jiuⅢ.Ji zhong lei xing sen lin qun luo de zhong-duo du guan xi de yan jiu. *Sheng tai xue bao*, 1997, **17**(6). 刘灿然, 马克平, 于顺利, 等. 北京东灵山地区植物群落多样性的研究Ⅲ.几种类型森林群落的种-多度关系研究. 生态学报, 1997, **17**(6)

Liu, C.R., Ma, K.P., Yu, S.L. deng. Bei jing dong ling shan di qu zhi wu qun luo duo yang xing de yan jiuⅣ.Yang ben da xiao dui duo yang xing ce du de ying xiang. *Sheng tai xue bao*, 1997, **17**(6). 刘灿然, 马克平, 于顺利, 等. 北京东灵山地区植物群落多样性的研究Ⅳ.样本大小对多样性测度的影响. 生态学报, 1997, **17**(6)

Liu, C.R., Ma, K.P., Yu, S.L. deng. Bei jing dong ling shan di qu zhi wu qun luo duo yang xing yan jiuⅥ.Ji zhong lei xing zhi wu qun luo wu zhong shu mu de gu ji. *Sheng tai xue bao*, 1998, **18**(2) 刘灿然, 马克平, 于顺利, 等. 北京东灵山地区植物群落多样性研究Ⅵ.几种类型植物群落物种数目的估计. 生态学报, 1998, **18**(2)

Liu, C.R., Ma, K.P., Yu, S.L. deng. Bei jing dong ling shan di qu zhi wu qun luo duo yang xing yan jiuⅦ.Ji zhong lei xing zhi wu qun luo lin jie chou yang mian ji de que ding. *Sheng tai xue bao*, 1998, **18**(1). 刘灿然, 马克平, 于顺利, 等. 北京东灵山地区植物群落多样性研究Ⅶ.几种类型植物群落临界抽样面积的确定. 生态学报, 1998, **18**(1)

Liu, C.R., Ma, K.P., Yu, S.L. deng. Bei jing dong ling shan di qu zhi wu qun luo duo yang xing yan jiu–Zhong-mian ji qu xian de ni he yu ping jia. *Zhi wu sheng tai xue bao*, 1999, **23**(6), 490–500. 刘灿然, 马克平, 于顺利, 等. 北京东灵山地区植物群落多样性研究一种-面积曲线的拟合与评价. 植物生态学报, 1999, **23**(6), 490–500

Liu, C.R., Ma, K.P., Zhou, W.N. Sheng wu qun luo duo yang xing de ce du fang fa ⅢYu wu zhong-Duo du fen bu mo xing you guan de tong ji wen ti. *Sheng wu duo yang xing*, 1995, **3**(3), 157–169. 刘灿然, 马克平, 周文能. 生物群落多样性的测度方法Ⅲ与物种－多度分布模型有关的统计问题. 生物多样性, 1995, **3**(3), 157–169

Liu, J., Chen, J.M., Cihlar, J., Park, W.M. A process-based boreal ecosystem productivity simulator using remote sensing inputs. *Remote Sensing of Environment*, 1997, **62**, 158–175

Liu, J.J. Da xing gong lu gong cheng sheng tai huan jing ying xiang de zong he ping jia. *Gan han qu zi yaun yu huan jing*, 1991, **5**(4), 34–42. 刘建军. 大型公路工程生态环境影响的综合评价. 干旱区资源与环境, 1991, **5**(4), 34–42

Liu, K.J., Liu, L., Zhou, C.X. Guan yu sheng wu duo yang xing zai lu yu zhi bei hui fu zhong de ying yong. *Jiao tong huan bao*, 2002, **16**(4). 刘孔杰, 刘龙, 周存秀. 关于生物多样性在路域植被恢复中的应用. 交通环保, 2002, **16**(4)

Liu, L. Wai lai zhi wu wu zhong de yin jin dui lu yu sheng tai de ying xiang. *Jiao tong huan bao*, 2003, **24**(2). 刘龙. 外来植物物种的引进对路域生态的影响. 交通环保, 2003, **24**(2)

Liu, L., Qian, D.S. Cong sheng tai xue jiao du hui fu gong lu zheng di fan wei de zhi bei. *Beijing: gong lu jiao tong ke ji*, 1999, **3**. 刘龙, 钱东升. 从生态学角度恢复公路征地范围的植被. 北京：公路交通科技, 1999, **3**

Liu, M.S., Zhang, M.J. *Jing guan sheng tai xue – yuan li yu fang fa*. Beijing: Hua xue gong ye chu ban she, 2004. 刘茂松, 张明娟. 景观生态学-原理与方法. 北京：化学工业出版社, 2004

Liu, R., Fang, X., Huang, Z.Y. Jiang su sheng di dai xing zhi bei de ji ben te dian ji fen bu gui lü. *Zhi wu sheng tai xue ji di zhi wu xue cong kan*, 1982, **6**(3). 刘日勋, 黄致远. 江苏省地带性植被的基本特点及分布规律. 植物生态学及地植物学丛刊, 1982, **6**(3)

Liu, S.Z., Gong, L.S. *Cheng shi jie dao lü hua she ji*. Beijing: Zhong guo jian zhu gong ye chu ban she, 1981. 刘少宗, 等. 城市街道绿化设计. 北京：中国建筑工业出版社, 1981

Ma, K.M., Ye, W.H., Sang, W.G. deng. Bei jing dong ling shan di qu zhi wu qun luo duo yang xing yan jiuⅩ.Bu tong chi du xia qun luo yang dai de β duo yang xing ji fen xing fen xi. *Sheng tai xue bao*, 1997, **17**(6). 马克明, 叶万辉, 桑卫国, 等. 北京东灵山地区植物群落多样性研究Ⅹ.不同尺度下群落样带的β多样性及分形分析. 生态学报, 1997, **17**(6)

Ma, K.P. Sheng wu qun luo duo yang xing de ce du fang fa Ⅰ αDuo yang xing de ce du fang fa (上). *Sheng wu duo yang xing*, 1994, **2**(3),162–168. 马克平. 生物群落多样性的测度方法Ⅰα多样性的测度方法(上). 生物多样性, 1994, **2**(3), 162–168

Ma, K.P., Liu, Y.M. Sheng wu qun luo duo yang xing de ce du fang fa Ⅰ αDuo yang xing de ce du fang fa (xia). *Sheng wu duo yang xing*, 1994, **2**(4), 231–239马克平, 刘玉明. 生物群落多样性的测度方法Ⅰα多样性的测度方法(下). 生物多样性, 1994, **2**(4), 231–239

Ma, K.P., Huang, J.H., Yu, S.L. deng. Bei jing dong ling shan di qu zhi wu qun luo duo yang xing de yan jiuⅡFeng fu du, jun yun du he wu zhong duo yang xing zhi shu. *Sheng tai xue bao*, 1995, **15**(3). 马克平, 黄建辉, 于顺利, 等. 北京东灵山地区植物群落多样性的研究Ⅱ丰富度、均匀度和物种多样性指数. 生态学报, 1995, **15**(3)

Ma, K.P., Liu, C.R., Liu, Y.M. Sheng wu qun luo duo yang xing de ce du fang fa II βDuo yang xing de ce du fang fa. *Sheng wu duo yang xing*, 1995, **3**(1), 38–43. 马克平, 刘灿然, 刘玉明. 生物群落多样性的测度方法II β多样性的测度方法. 生物多伴性, 1995, **3**(1), 38–43

Mao, L.S. *Guan shang shu mu xue*. Nanjing: Nan jing nong ye da xue, 1995. 毛龙生. 观赏树木学. 南京 : 南京农业大学, 1995

Matsushita, B., Tamura, M. Integrating remotely sensed data with an ecosystem model to estimate net primary productivity in East Asia. *Remote Sensing of Environment*, 2002, **81**, 58–66

McHarg, I.L. *She ji jie he zi ran. Rui Jingwei yi*. Beijing: Zhong guo jian zhu gong ye chu ban she, 1992. 麦客哈格IL. 设计结合自然. 芮经维译. 北京 : 中国建筑工业出版社, 1992

Meng, Q. Qing hao sheng ping an zhi xi ning gao su gong lu jing guan lü hua she ji. *Gong lu jiao tong ke ji*, 2001, **18**(5). 孟强等. 青海省平安至西宁高速公路景观绿化设计.公路交通科技, 2001, **18**(5)

Pan, L., Liu, S.S., Lü, H.Y. Wu da lian chi xin qi huo shan de zhi bei lei xing ji qun luo te zheng de yan jiu. *Sheng tai xue za zhi*, 1998, **15**(81). 潘林, 刘士山, 吕洪彦. 五大连池新期火山的植被类型及群落特征的研究. 生态学杂志, 1998, **15**(81)

Patil, G.P., Taillie, C. Diversity as a concept and its measurement. *American Statistical Association*, 1982, **77**, 548–567

Peng, H., Hou, L., Yang, K.N. Gao su gong lu jian she dui qu yu huan jing de ying xiang yu jing guan xi tong fen xi fang fa. *Zhong guo shui tu bao chi*, 2003, **5**, 16–17. 彭鸿, 候琳, 杨康宁. 高速公路建设对区域环境的影响与景观系统分析方法. 中国水土保持, 2003, **5**, 16–17

Qi, R. Gao su gong lu jian she dui huan jing de ying xiang ji bao hu cuo shi. *Shan xi jiao tong ke ji*, 2003, **2**. 齐荣. 高速公路建设对环境的影响及保护措施. 山西交通科技, 2003, **2**

Rapoport, A. *Jian cheng huan jing de yi yi-Fei yan yu biao da fang fa*. Beijing: Zhong guo jian zhu gong ye chu ban she, 2003. [美]阿摩斯·拉普卜特. 建成环境的意义－非言语表达方法. 北京 : 中国建筑工业出版社, 2003

Rosenzweig, M.L., Sandlin, E.A. Species diversity and latitudes: listening to area's signal. *Oikos*, 1997, **80**, 172–176

Shan, B.H. *Qiao Liang mei xue*. Beijing: Ren min jiao tong chu ban she, 1987. [日]山本宏. 桥梁美学. 北京 : 人民交通出版社, 1987

Shen, T.G., Shen,X.Y., Liu, X. deng. *Qiao liang jian zhu yu xiao pin-gou si yu zao xing*. Tianjin: Tian jin da xue chu ban she, 2002. 慎铁刚, 等. 桥梁建筑与小品－构思与造型. 天津 : 天津大学出版社, 2002

Shen, Z.H., Jin, Y.X., Zhao, Z.E. Ya re dai shan di sen lin zhen xi zhi wu qun luo de jie gou yu dong tai. *Sheng tai xue bao*, 2000, **20**(5). 沈泽昊, 金义兴, 赵子恩. 亚热带山地森林珍稀植物群落的结构与动态. 生态学报, 2000, **20**(5)

Sheng, H.F. *Qiao liang jian zhu mei xue*. Beijing: Ren min jiao tong chu ban she, 1999. 盛洪飞. 桥梁建筑美学. 北京 : 人民交通出版社, 1999

Sheng, Y. *Qiao liang mei xue zhong guan yu jing guan she ji yu ping jia de yan jiu*. Xi'an: Chang an da xue shuo shi lun wen, 2001. 盛勇. 桥梁美学中关于景观设计与评价的研究. 西安 : 长安大学硕士论文, 2001

Shi, L.M., Luo, D.C. Gong lu sheng tai huan jing ying xiang ping jia de ji zhong ding liang fen xi fang fa. *Zhong guo gong lu xue bao*, 1998, **11**(1), 43–53. 师利明, 罗德春. 公路生态环境影响评价的几种定量分析方法. 中国公路学报, 1998, **11**(1), 43–53

Shu, X., Du, J., Cao, Y.H. deng. Sheng tai gong cheng zai gao su gong lu yan shi bian po fang hu gong cheng zhong de ying yong. *Gong lu*, 2001, **7**. 舒翔, 等. 生态工程在高速公路岩石边坡防护工程中的应用. 公路, 2001, **7**

Su, Z.Y., Chen, B.G., Gu, Y.K. deng. Guang zhou bai yun shan ji zhong sen lin qun luo de wu zhong feng fu du he duo yang xing. *Guangzhou: hua nan nong ye da xue xue bao*, 2001, **22**(3). 苏志尧,陈北光,古炎坤,等. 广州白云山几种森林群落的物种丰富度和多样性. 广州 : 华南农业大学学报, 2001, **22**(3)

Su, Z.Y., Chen, B.G., Wu, D.R. Guang dong ying de shi men tai zi ran bao hu qu de zhi bei lei xing he qun luo jie gou. *Guangzhou: hua nan nong ye da xue xue bao*, 2002, **23**(1). 苏志尧,陈北光,吴大荣. 广东英德石门台自然保护区的植被类型和群落结构. 广州 : 华南农业大学学报, 2002, **23**(1)

Sun, J. *Qiao liang huan jing yi shu de li lun tan tao yu shi jian yan jiu*. Wuhan: Wu han li gong da xue shuo shi lun wen, 2001. 孙菁. 桥梁环境艺术的理论探讨与实践研究. 武汉 : 武汉理工大学硕士论文, 2001

Sun, Q.B., Zhen, X.Y. Gao su gong lu jian she dui sheng tai huan jing de ying xiang ji hui fu. *Kunming: kunming li gong da xue xue bao*, 2000, **25**(2). 孙乔宝,甄晓云. 高速公路建设对生态环境的影响及恢复.昆明 : 昆明理工大学学报, 2000, **25**(2)

Tan, H.C. *Qiao*. Beijing: Zhong guo tie dao chu ban she, 1981. 唐寰澄. 桥.北京 : 中国铁道出版社, 1981

Turner, M.G. Spatial and temporal analysis of landscape patterns. *Landscape Ecology*, 1990, **4**, 21–30

Turner, M.G., Gardner, R.H. *Quantitative Methods in Landscape Ecology*. New York: Springer-Verlag, 1991

Wang, P. *Huan jing yi shu she ji*. Beijing: Zhong guo fang zhi chu ban she, 1998. 王朋. 环境艺术设计.北京 : 中国纺织出版社, 1998

Wang, Y.A., Wang, S.S. Gong lu lü di xi tong de sheng tai xue fen xi. *Hua dong gong lu*, 2002, **3**. 王永安,王双生. 公路绿地系统的生态学分析. 华东公路, 2002, **3**

Wang, Z.Y. *Qiao liang man hua*. Beijing: Ren min jiao tong chu ban she, 1994. 王展意.桥梁漫话.北京 : 人民交通出版社, 1994

Wei, F.H., Chen, H., Wang, Z.Y. Gao su gong lu jian she qi sheng tai huan jing ying xiang fen xi. *Liao ning sheng jiao tong gao deng zhuan ke xue xiao xue bao*, 2003, **5**(2). 魏凤虎,陈　红,王卓娅.高速公路建设期生态环境影响分析.辽宁省交通高等专科学校学报, 2003, **5**(2)

Whittaker, R.H. Evolution and measurement of species diversity. *Taxon*, 1972, **21**, 213–251

Whittaker, R.H. Gradient analysis of vegetation. *Biological Reviews*, 1967, **49**, 207–264

Wilson, J.B. Methods for fitting dominance/diversity curves. *Journal of Vegetation Science*, 1991, **2**, 35–46.

Wilson, M.V., Shmida, A. Measuring beta diversity with presence-absence data. *Journal of. Ecology*, 1984, **72**, 1055–1062

Wolf, K.L. Freeway roadside management: the urban forest beyond the white line. *Journal of Arboriculture*, 2003, **29**(3), 127–136

Wu, J.G. *Jin guan sheng tai xue –ge ju, guo cheng, chi du yu deng ji*. Beijing: Gao deng jiao yu chu ban she, 2000. 邬建国. 景观生态学-格局、过程、尺度与等级.北京 : 高等教育出版社, 2000

Wu, Y., Su, Z.X. Lu di zhi wu qun luo wu zhong duo yang xing yan ti yan jiu jin zhan. *Sheng ming ke xue yan jiu*, 2001, **5**(3). 吴勇,苏智先.陆地植物群落物种多样性演替研究进展. 生命科学研究, 2001, **5**(3)

Xi, J.B., Zhang, H.X. Ji zhong hun bo lü hua zu he dui gao deng ji gong lu bian po fang hu xiao yi de yan jiu. *Cao ye ke xue*, 2000, **17**(4). 席嘉宾,张惠霞. 几种混播绿化组合对高等级公路边坡防护效益的研究. 草业科学, 2000, **17**(4)

Xia, H.R. Gao su gong lu huan jing jing guan ping jia de yan jiu. *Huan jing bao hu ke xue*, 2001, **27**, 42–43. 夏惠荣. 高速公路环境景观评价的研究.环境保护科学, 2001, **27**, 42–43

Xiang, W.D., Guo, J., Wei, Y., Zhang, J.C. Gao su gong lu jian she dui qu yu sheng wu duo yang xing ying xiang de ping jia. *Nan jing lin ye da xue xue bao (zi ran ke xue ban)*, 2003, 27(6), 43–47. 项卫东, 郭建, 魏勇, 张金池. 高速公路建设对区域生物多样性影响的评价.南京林业大学学报(自然科学版), 2003, 27(6), 43–47

Xiao, D.N., Bu, R.C., Li, X.Z. *Shi lun jing guan sheng tai xue de li lun ji chu yu fang fa lun te dian. Xiao, D.N. Jing guan sheng tai xue li lun, fang fa ji qi ying yong.* Beijing: Zhong guo lin ye chu ban she, 1991. 肖笃宁, 布仁仓, 李秀珍. 试论景观生态学的理论基础与方法论特点.肖笃宁.景观生态学理论、方法及其应用.北京 : 中国林业出版社, 1991

Xiong, G.Z. *Cheng shi dao lu mei xue.* Beijing: Zhong guo jian zhu gong ye chu ban she, 1990. 熊广忠. 城市道路美学.北京 : 中国建筑工业出版社, 1990

Xiong, G.Z. Gong lu lü hua chu bu tan tao. *Zhong guo gong lu xue bao*, 1995, 8(4). 熊广忠.公路绿化初步探讨.中国公路学报, 1995, 8(4)

Xu, H.C. *Jing guan sheng tai xue.* Beijing: Zhong guo lin ye chu ban she, 1996. 徐化成. 景观生态学.北京 : 中国林业出版社, 1996

Yang, F.J., Zhao, Z.H., Fu, Y.J. deng. Feng shan yu lin hou tian ran ci sheng lin qun luo jie gou te zheng. *Zhi wu yan jiu*, 2002, 22(4). 杨逢建, 赵则海, 付玉杰, 等. 封山育林后天然次生林群落结构特征.植物研究, 2002, 22(4)

Yang, S.J., Tang, H.X. *Jing guan qiao liang she ji.* Shanghai: Tong ji da xue chu ban she, 2003. 杨士金, 等. 景观桥梁设计.上海 : 同济大学出版社, 2003

Ye, W.H., Ma, K.P., Ma, K.M. deng. Bei jing dong ling shan di qu zhi wu qun luo duo yang xing yan jiuIX.Chi du bian hau dui α duo yang xing de ying xiang. *Sheng tai xue bao*, 1998, 18(1). 叶万辉, 马克平, 马克明, 等. 北京东灵山地区植物群落多样性研究IX.尺度变化对α多样性的影响.生态学报, 1998, 18(1)

Yi, T.X. *Qiao liang zao xing.* Beijing: Ren min jiao tong chu ban she, 1998. 伊藤学.桥梁造型.北京 : 人民交通出版社, 1998

You, M.S. Qun luo duo yang xing de yan jiu jin zhan (Ying wen). *Fuzhou: fu jian nong ye da xue xue bao*, 1998, 27(4). 尤民生. 群落多样性的研究进展(英文). 福州 : 福建农业大学学报, 1998, 27(4)

Yu, D.R. Qiao liang mei xue man tan. *Guo wai gong lu*, 1998, 18(1). 余丹如. 桥梁美学漫谈.国外公路, 1998, 18(1)

Yu, K.J. *Jing Guan: Wen hua, Sheng tai yu gan zhi.* Beijing: Ke xue chu ban she, 1998. 俞孔坚. 景观 : 文化, 生态与感知.北京 : 科学出版社, 1998

Yu, S.X., Zhang, H.D., Wang, B.S. Hai nan dao ba wang ling re dai shan di zhi bei yan jiuII : Dai biao qun luo lei xing de jie gou fen xi. *Sheng tai ke xue*, 1994, 1. 余世孝, 张宏达, 王伯荪. 海南岛霸王岭热带山地植被研究II : 代表群落类型的结构分析.生态科学, 1994, 1

Yu, Z.L. *Cheng shi huan jing yi shu.* Tianjin: Tian jin ke xue ji shu chu ban she, 1990. 于正伦.城市环境艺术.天津 : 天津科学技术出版社, 1990

Yuan, C.M., Lang, N.J., Meng, G.T. deng. Chang jiang shang you hua shan song shui tu bao chi ren gong luo de jie gou te zheng yu sheng wu liang. *Shenyang: dong bei lin ye da xue xue bao*, 2002, 20(3). 袁春明, 郎南军, 孟广涛, 等. 长江上游华山松水土保持人工群落的结构特征与生物量.沈阳 : 东北林业大学学报, 2002, 20(3)

Yuan, G.L. *Gong lu yu huan jing jing guan she ji.* Nanjing: Dong nan da xue shuo shi lun wen, 2001. 袁国林. 公路与环境景观设计.南京 : 东南大学硕士论文, 2001

Zhang, F.G. Zhe jiang qing liang feng tai wan shui qing gang lin de qun luo xue te zheng. *Zhe jiang da xue xue bao (nong ye yu sheng ming ke xue ban)*, 2001, 27(4). 张方钢.浙江清凉峰台湾水青冈林的群落学特征.浙江大学学报(农业与生命科学版), 2001, 27(4)

Zhang, J.E., Xu, Q. Guan yu dao lu de sheng tai xue ying xiang ji qi sheng tai jian she. *Sheng tai xue za zhi*, 1995, **14**(6). 章家恩, 徐琪. 关于道路的生态学影响及其生态建设. 生态学杂志, 1995, **14**(6)

Zhang, L., Lin, W.Q., Chen, B.G. deng. Guang zhou mao feng shan ci sheng lin qun luo jie gou te zheng. *Hua nan nong ye da xue xue bao*, 2003, **24**(3). 张璐, 林伟强, 陈北光, 等. 广州帽峰山次生林群落结构特征. 华南农业大学学报, 2003, **24**(3)

Zhang, P.F., Yao, C. Gao su gong lu yu cheng shi dao lu yan xian jiao tong dui huan jing de wu ran ji zhi li cuo shi qian xi. *Cheng shi dao qiao yu fang hong*, 2001, **2**. 张鹏飞, 姚成. 高速公路与城市道路沿线交通对环境的污染及治理措施浅析. 城市道桥与防洪, 2001, **2**

Zhang, P.F., Yao, C. Gao su gong lu yu cheng shi dao lu yan xian jiao tong zao sheng dui huan jing de wu ran fen xi. *Chng shi huan jing yu cheng shi sheng tai*, 1999, **26**(3). 张鹏飞, 姚成. 高速公路与城市道路沿线交通噪声对环境的污染分析. 城市环境与城市生态, 1999, **26**(3)

Zhang, S.D. *Qiao liang jian zhu de jie gou gou si yu she ji ji qiao*. Beijing: Ren min jiao tong chu ban she, 2002. 张师定. 桥梁建筑的结构构思与设计技巧. 北京 : 人民交通出版社, 2002

Zhang, S.E. Ning xia gu wang gao su gong lu jian she zhong sheng tai huan jing bao hu cuo shi. *Ning xia nong xue yuan xue bao*, 2003, **24**(1). 张淑娥. 宁夏古王高速公路建设中生态环境保护措施. 宁夏农学院学报, 2003, **24**(1)

Zhang, S.S. Gao deng ji gong lu bian po de sheng wu fang hu ji shu. *Gong lu*, 2002, **9**. 张世绥. 高等级公路边坡的生物防护技术. 公路, 2002, **9**

Zhang, Y., Yuan, W.N., Li, P. Lu qiao gong cheng zhong dang tu qiang jing guan she ji fang fa yan jiu. *Xi bei jian zhu gong cheng xue yuan xue bao*, 2000, **17**(3). 张阳, 等. 路桥工程中挡土墙景观设计方法研究, 西北建筑工程学院学报, 2000, **17**(3)

Zhang, Y.F. Gao deng ji gong lu jian she yu huan jing xie tiao fa zhan ruo gan wen ti de tan tao. *Jiao tong huan bao*, 1999, **20**(1). 张玉芬. 高等级公路建设与环境协调发展若干问题的探讨. 交通环保, 1999, **20**(1)

Zhao, Y., Sun, Z.D., Wu, M.Z. Gao su gong lu jian she xiang mu dui sheng tai huan jing ying xiang zong he ping jia yan jiu. *Anquanyuhuanjinggongchen*, 2003, **10**(3), 12–15. 赵勇, 孙中党, 吴明作. 高速公路建设项目对生态环境影响综合评价研究. 安全与环境工程, 2003, **10**(3), 12–15

Zhou, H.C., Peng, S.L., Ren, H. Guang dong nan ao lü ma wei song lin de qun luo jie gou. *Re dai ya re dai zhi wu xue bao*, 1998, **6**(3). 周厚诚, 彭少麟, 任海. 广东南澳吕马尾松林的群落结构. 热带亚热带植物学报, 1998, **6**(3)

Zou, D.Y. *Jian zhu xing shi mei de yuan ze*. Beijing: Zhong guo jian zhu gong ye chu ban she, 1992 邹德侬. 建筑形式美的原则. 北京 : 中国建筑工业出版社, 1992

Index

The Environment and Landscape in Motorway Design, First Edition.
Qian Guochao, Tang Shuyu, Zhao Min and Jing Chun.
© 2014 China Communications Press. Published 2014 by John Wiley & Sons, Ltd.